Routledge Revivals

The Rhythms of Life

Originally published in 1929, this title is a series of essays on science at the beginning of the twentieth century. Including biology, chemistry, physics and psychology it covers a range of topics, from heating the house to suspended animation. This title can now be read and enjoyed in its historical context.

The Rhythms of Life

and Other Essays in Science

D. F. Fraser-Harris

Routledge
Taylor & Francis Group

First published in 1929
by George Routledge & Sons Ltd.

This edition first published in 2021 by Routledge
2 Park Square, Milton Park, Abingdon, Oxon OX14 4RN

and by Routledge
605 Third Avenue, New York, NY 10017

Routledge is an imprint of the Taylor & Francis Group, an informa business

Publisher's Note
The publisher has gone to great lengths to ensure the quality of this reprint but points out that some imperfections in the original copies may be apparent.

Disclaimer
The publisher has made every effort to trace copyright holders and welcomes correspondence from those they have been unable to contact.

ISBN: 978-1-032-01179-0 (hbk)
ISBN: 978-1-003-17763-0 (ebk)

THE RHYTHMS OF LIFE

AND OTHER ESSAYS IN SCIENCE

By
D. F. FRASER-HARRIS
M.D., D.Sc., F.R.S.E.

LONDON
GEORGE ROUTLEDGE & SONS LTD.
BROADWAY HOUSE: 68-74 CARTER LANE, E.C.
1929

PRINTED IN GREAT BRITAIN BY
M. F. ROBINSON & CO. LTD., AT THE LIBRARY PRESS, LOWESTOFT

PREFACE

Some of the essays included in this volume have already appeared either on this or the other side of the Atlantic.

It is a pleasure to thank the Editor of *The Cornhill Magazine* for permission to republish the article dealing with Plague and the Litany; and the Editor of *Chambers' Journal* for his kindness in allowing me to print again the articles on "Suspended Animation" and on "Blaming the Air". I have also to express my indebtedness to the Editor of *Nature* for granting me permission to use the article on "Physiology and Vital Force" which was published in that journal in April, 1925; and to the Editor of *Discovery* for allowing me to reprint the articles on "Animal Electricity" and "Animal Heat" which appeared in *Discovery*.

Through the kindness of the Editor of *The Birmingham Mail*, I am enabled to reprint the article on " What the microscope has achieved for mankind ", which was originally published in the *Mail* in November, 1927. The essay on " Lost Arts " first appeared in 1925 in the July number of the *Dalhousie Review*, and I have to thank my former colleague, Professor H. L. Stewart, the Editor, for his courtesy in enabling it to reappear as a chapter of this book.

D. F. F.-H.

CHISWICK.
January, 1929.

CONTENTS

THE RHYTHMS OF LIFE

CHAPTER I

A PHENOMENON is called rhythmic when it recurs at equal intervals of time, or, at any rate, not at cognisably unequal ones. The dripping of water from a leaky tap is rhythmic, but the noise of a tray falling downstairs is not. The universe is full of rhythms. The succession of the seasons, the alternation of day and night, the phases of the moon, the ebb and flow of the tide, the recurrence of spring and neap tides, the November flight of meteors, the yearly rise of the Nile, and so forth are all examples of cosmic rhythms. The precise magnitude of the time-interval is of no consequence, for in the case of waves of light the interval may be only some millionths of a second, whereas the return of a comet, such as Halley's, to our solar system, may be a matter of seventy years or so. The vibrations of the ether and the behaviour of the comet are both rhythmical or periodic.

Music is essentially rhythmic; in fact, it is the periodic character of the vibrations of the air that constitutes music as distinguished from noise. The vibrations of the air objectively constituting noise are highly irregular or arhythmic.

A clap of thunder is not rhythmical. In a rhythm something occurs at equal intervals of time; if this something recurs at unequal intervals, the rhythm sometimes is spoken of as "irregular." The ordinary usage is to make the word "rhythm" synonymous with "regular rhythm", the only sense in which the term will be employed in what follows. It is owing to the regularity of recurrence of eclipses, the equinox, meteors, comets, etc., that these phenomena can be predicted with an accuracy that makes astronomy as an exact science so justly admired. The periodicities of rhythms in the non-living world may be matters of years, months, weeks, days, hours, minutes, seconds or fractions of a second.

Coming now to the realm of life, we find rhythms pervading everything. The plants, with striking regularity, have their own times each year for putting forth the buds, unfolding the leaves, bursting into flower, and finally, allowing all the perfumed beauty of the corollas to fade in order that the fruit shall be formed as a life in death. Thus the poetess sang—

> "Leaves have their times to fall,
> And flowers to wither at the north wind's breath;
> Thou hast all seasons for thine own, O Death."

In plain prose, there is no rhythm about death.

"Chestnut Sunday" is approximately the same Sunday, and "Apple Blossom" week is as near as may be the same week, each year.

If one lives in the country one is able to watch the unerring precision with which each kind of tree puts forth its blossom. The wild plum or sloe is one of the earliest, then comes the pear, then the cherry, then the apple, and so on. Each has its own week for the annual exhibition of virginal loveliness.

The opening and closing of flowers is rhythmic, the rhythm depending on the waxing and waning of the intensity of daylight.

Doubtless the most familiar rhythms are met with in the world of animal life. Here we have the striking rhythmic actions of animals as in flocks and herds, of animals as individuals, and of the organs, tissues and cells of the animal body. The migration of birds is annual in rhythm. It is only part of the truth to say that the waning light and diminished heat and food are what constrain the birds to leave us; they *know* when to leave us. Migratory birds which have spent all their lives well-fed in the captivity of Regent's Park, nevertheless become restless at the approach of autumn. Again, those animals which hibernate during the winter know when to betake themselves to their hiding places, whence they come not forth until the spring. The cuckoo returns almost on the same day every year.

The rhythm of the sexual activities of birds is one of the most characteristic things about their behaviour. It is only in the spring that the "livelier iris changes on the burnish'd dove", but it is *every* spring. Even a decerebrated male pigeon will "coo" energetically at the breeding season, although if the hen bird be placed near him he will take no notice of her. The sexual rhythm is inherent in the nervous system, but in the absence of the brain it is a meaningless and mechanical rhythm.

Only those who have lived much in the country have any adequate notion of the reality of these biological rhythms.

Practically all the activities of one's daily life are rhythmic; the most obvious perhaps being the regular alternation of waking and sleeping. There is a well-marked rhythm in our digestive organs, in the organs of excretion, and most pronouncedly in the beating heart.

Rhythm pervades the world of animal life; just watch that transparent jelly-fish in the limpid summer sea, and you will notice how the edges of the umbrella contract or pulsate with that slow and regular rhythm of about thirty in the minute. Equally obvious rhythms are those of the wings of birds and other flying things, of fins in swimming, of the legs in walking and dancing. Large birds fly with slow, leisurely rhythm, small birds with a fast one; just as tall men have a slow stride, short men have a more rapid step. Regular rhythms are everywhere; if Nature abhors a vacuum, she also abhors "fits and starts"; living Nature does everything "decently and in order".

The periodicity of the heart's action is an

excellent example of a rhythm of animal origin. Seventy-two times in the minute this wonderful hollow muscle contracts in systole on the contained blood forcing it out into all the arteries of the body; and seventy-two times in the minute does it dilate in diastole and suck blood in from the veins leading to it. The duration of its cycle is therefore eight-tenths of a second ($\frac{60}{72}$); and in health during the many years of a long life it practically does not vary from this value. If for more than a few minutes the heart's rhythm remains distinctly irregular, then we conclude that something is amiss with this wonderful organ within the breast.

Sometimes we come across a heart with a congenitally fast rhythm, a condition called tachycardia; and sometimes one with an abnormally slow rhythm, a condition called bradycardia. Whereas the rhythm of the heart-beat is for each individual one of a certain average rate, it varies in different individuals according to height and age. Thus, tall persons have slower hearts than short people; and infants have a heart-rate about twice as fast as adults. The whale and the elephant have very much slower heart-beats than the mouse or the sparrow. The heart-beat in these small animals is so fast that the pulse cannot be counted in the usual way. Until recently we had no reliable information about it, but by an electrical method it has been ascertained with great accuracy. For instance, the rabbit's heart beats at four a second.

The pulse-rate, which in health is the same thing as the heart-rate, is very much slower in the cold-blooded than in the warm-blooded animals. Thus, in a fish or frog the heart contracts only about forty times in the minute, or about half the human rate. This relatively slow rate can be quickened by making the heart beat in warm salt water. A study of the warm-blooded heart proves to be full of interest. The heart can be accelerated by warmth and slowed by cold, but it is its affectability by nerve impulses that is so remarkable. It is a matter of common knowledge that the heart can be made to beat much faster at one time and slower at another through nerve impulses alone. Everybody knows that emotions can influence the heart very markedly. Thus, of course, it has come about that "the heart" and certain emotions are taken as synonymous. Certain emotions disturb it in that they cause it to beat more rapidly, while others slow it and enfeeble it.

Now physiologists have discovered two distinct sets of nerves which influence the heart-rate; the one set on being stimulated makes the heart faster and stronger (accelerator and augmentor nerves), the other set on being stimulated makes the heart beat more slowly and may stop it altogether (inhibitory nerves). Evidently the former set arouses the heart to greater activity, the latter induces in it less activity, than the normal.

The rhythmicality of the heart is not conferred on it by the action of nerves, by the presence of blood, by the temperature of the blood, or by any other "external" condition; its rhythmicality is inherent, it is spontaneous or autogenic. The rhythm of the heart is of the essence of its life; the microscopic cells of the embryo heart beat with a definite rhythm as soon as they are perceptible at all, and before nerves have reached them or the blood has been formed.

The heart is not the only rhythmic portion of the circulatory system. In all animals, portions of the large veins have the power of rhythmic contraction : in the bat they contract sixteen times per minute. In some animals (frogs, for instance) there are pulsatile sacs or lymph-hearts, dilatations of the lymphatic vessels, beating visibly under the skin of the back.

Many other organs exhibit rhythms. The activity of the stomach is rhythmic, and that of the intestines over which waves of contraction pass at short intervals ; the activities also of the gall-bladder, the urinary bladder and the uterus are all rhythmic.

An interesting thing about rhythmic organs is their inability to have the rate of their rhythm forced beyond a certain limit. No amount of stimulation of the accelerator nerves can increase the rate of the heart-beat beyond a certain limit. Similarly, heating the heart will raise the rate of the rhythm, but only up to a particular figure which cannot be exceeded. Working with the pigeon, the author found that the greatest number of beats per minute which the heart (auricle) could give was 300, or five a second, and beyond that it was not possible to force it. Not only, then, is rhythmicality inherent in the living substance of the heart, but also a power of resisting all influences to accelerate it beyond a certain limit, an expression of what has been called "functional" or "physiological inertia".

Let us take another example of rhythmic activity as seen in the cilium. A cilium is a minute, whip-like outgrowth of living protoplasm projecting from the surface of a cell. There are millions of these cilia covering the mucous membranes of the nose, throat and bronchial tubes, which lash mucus, dust and germs towards the nostrils and mouth. Now, these cilia lash backwards and forwards at a characteristic rate of ten to twelve in the second. Just as the heart makes seventy-two systoles and seventy-two diastoles in a minute, so the cilium bends forwards ten times and backwards ten times in a second, or 600 times a minute, about eight times as fast as the heart. Otherwise put, the period of ciliary oscillation is one-tenth of a second instead of eight-tenths.

But the rhythmicality of the cilium is as inherent as that of the heart. The cilia receive no nerves; therefore, not being innervated, they cannot possess any rhythm conferred on them from outside by nerves, the neurogenic rhythm. The cilia are, however, easily influenced in their rate of vibration by changes of temperature and by drugs and poisons. By warming the cilia they lash faster and faster until they attain a speed of about twenty a second, beyond which they cannot go, one more example of a limit set. Conversely, cold and narcotics like chloroform slow and finally stop their action just as they do that of the heart.

Let us take the case of breathing. Normally, an adult breathes about sixteen to eighteen times a minute, that is to say, in a wholly unconscious fashion his diaphragm—the great, curved muscle between the chest and the abdomen—descends eighteen times and rises eighteen times in the minute. There is, therefore, a respiratory rhythm just as there is a cardiac and a ciliary. Now the diaphragm would not make any descents were it not that it was receiving nerve-impulses through its nerves (phrenics). After these nerves have been cut, the diaphragm is absolutely still. Clearly, then, the rhythm of the activity of the diaphragm is not inherent, but, on the contrary, is conferred by nerves or is neurogenic. The rhythm of eighteen to twenty a second must be the rhythm of discharge of nerve impulses from the nerve cells or " centres " from which the phrenic nerves come. It is the nerve cells that have this rhythm, not the nerves as conductors and not the diaphragm as a muscle.

The actual cells from which the phrenic nerves proceed are, however, not the breathing centre, which, situated higher up in the central nervous system, constrains the phrenic centres to issue their periodic discharges of nerve-impulses. The chief respiratory centre possesses the real respiratory rhythm, which, like the heart's, varies with age and other circumstances. We know very well that the rate of breathing can be profoundly altered by emotional states: some conditions —for instance, states of excitement—greatly increase the rate; others slow it or stop it for a time, as in the phrase, "it fairly took my breath away".

Experiments have shown that heated blood accelerates the breathing and that cold slows it, and there are drugs which have analogous actions. Further, the will can for a time abolish the rhythm altogether. Divers are able intentionally to "hold their breath": on the other hand, the will, if we so wish it, can hurry up the rhythm beyond the rate of the normal. This property of having a rhythm which can be altered by the will is quite a rare one amongst the centres of the nervous system.

The normal respiratory rhythm is, then, an additional example of a rhythm inherent in something—in this case in the cells of a nerve centre—but capable of responding to outside influences. And again, there are limits set, for neither by the will, nor by emotion, nor by heated blood, nor by drugs, can the rate of the breathing centre be forced beyond a certain maximum value.

Breathing is to all intents and purposes an involuntary, unconscious activity; the diaphragm rises and falls throughout life, whether we wake or whether we sleep, with a regularity that is as constant and with efforts that are as untiring as those of the heart itself. In a word, the rhythm of breathing is not voluntary, although we can interfere with it voluntarily.

But, of course, the nervous system can give us many examples of rhythms of voluntary origin. Take the very simple case of tapping one's finger on the table or on an electric key. I can tap my forefinger once a second, twice a second, three times a second, and so on, until I am tapping it so fast I can barely count it. When this rate is reached, an instrument of simple construction can prove that the finger is being flexed and extended at from ten to twelve times a second. We may note in passing that this is exactly the same as the ciliary rate. Now the instrument will show that beyond ten to twelve a second the ordinary person cannot go, although it is possible to train musical technicians to "trill" at a considerably greater rate than that of ordinary people. Still, even for experts a limit is soon reached. The rhythm of the cells of the centres giving rise to the nerves to the fingers is evidently of such a kind, that, whereas the cells can be made by the will to assume any slow rhythm from one to twelve a second, they cannot ordinarily be forced beyond their limit.

What, then, is the rate of inherent rhythm of these centres? Probably ten to twelve a second. It is reasonable to suppose that these cells have a normal, natural, inherent rate of discharge which may be the same as their maximal rate. Neither by the will nor by artificial stimulation can this maximum be exceeded, not even when the rhythm of the artificial stimulation is much higher than twelve a second. The nerve cells have physiological inertia towards rates of stimulation greater than that of their own maximal rhythm. The respiratory centre, on the contrary, we saw, had a normal rate which was *not* also its maximal.

Rhythm or intermittency pervades the nervous system. It has been ascertained that we cannot utter syllables (articulate) at a greater rate than ten to twelve a second. What we may call the articulation centre has its upper limit set at this figure which we have so frequently encountered already. The numerical identity of the rhythms of cilium, musculo-motor nerve centre and articulation centre is probably not accidental.

The frequencies of insects' muscles are of the following orders: The wings of the dragon-fly vibrate at twenty-eight a second, those of the wasp at 110, of the bee at 190, and of the house-fly at 330 per second. The late Professor Mosso asserted that the pitch of the note of a bee setting out on its day's rounds was perceptibly higher than that of the note heard at the close of the day.

Finally, it is probable that the receiving or sensory portions of the brain are constructed in such a manner that they, too, have limits in dealing with rhythmic or intermittent presentations.

The spokes of a slowly rotating bicycle wheel can be perceived as separate bright lines, but when the wheel is revolving rapidly the individual spokes fuse into one bright metallic surface, just as the separate slats of a paling viewed from an express train fuse into one continuous surface. The grooves of the milling on the edge of a metal disc spun rapidly under the finger are perceived as constituting a rough but continuous surface. The fusion of the members of a series of instantaneous photographs of moving objects presented in very rapid succession to the eye, as in the kinematograph, is due to this incapacity of the brain to resolve as distinct in consciousness the separate components of the physical series.

If the interval between the successive impressions is longer than about one-fortieth to one-fiftieth of a second we do not get fusion, but the well-known and disagreeable state of " flicker ". This was common in the early moving-picture houses, called incorrectly " cinemas ". These and many other cases prove that there are strict limits to the perception of rhythms by our brains.

The causes of vital rhythms and periodicities are virtually unknown. Physiologists can describe vital rhythmic actions in their own precise language, but that is all. To say that the cardiac cycle is constituted by the two alternating and opposite phases of katabolic systole and anabolic diastole does not bring us much nearer an understanding of the cardiac rhythm. Protoplasm, in general, tends to act intermittently. Just as a single tap given to a jelly or to a spring will make these oscillate or vibrate for some considerable time thereafter, so a single or continued stimulus given to living matter will cause it to discharge energy in a vibratory or intermittent manner.

Nor do we comprehend any better the significance of the particular time-duration of the interval between the recurring events which are being studied. Why, for instance, should the human heart beat seventy-two times a minute and not seven or 700 ? Why should we breathe eighteen times and not eight or eighty times a minute ? Why should the cilium bend ten times a second as its normal rate and just twice as fast, but no more than that, when urged to do its utmost ? These things are mysteries. The rhythms of functional activity in the female reproductive organs are familiar facts of physiology, but what induces the rhythm and why the periodicity should be as it is are problems at present entirely unsolved.

Probably the necessity of rest to prevent fatigue or exhaustion is one of the purposes of vital rhythms. The heart, for instance, can continue to beat so indefatigably just because the duration of its time of rest (diastole) in the cycle is longer than that of its activity (systole). We sleep by night in order to be active by day. All work and no rest is a physiological outrage ; rhythm is an expression of that physiological normality in which work alternating with rest is most economically performed.

Spontaneous rhythmicality is the great mystery of life, the central puzzle in biology ; if we knew what rhythm really was, understood it "all in all", we should perhaps "know what God and man is".

CHAPTER II

ANIMAL ELECTRICITY

MANIFESTATIONS of electricity seem to be associated with almost every kind of activity in the universe. In the non-living world, such things as friction, the evaporation of water, the transference of heat, the passage of liquids through membranes, chemical activity and magnetism are all accompanied by the production of electricity. It would be difficult to name any physical process in which electricity does not participate directly or indirectly.

The living body and its tissues are no exception to this universal happening.

Let us first consider a muscle in the living body as regards its relations to energy. Since muscles are warm, they must be continually producing heat.

In fact, by far the greatest amount of heat set free in the body has had its origin in the muscles. Even when the muscles are what we call "at rest" they are still producing heat, though, of course, far less of it than when they are in action.

When muscles go into activity they evolve in addition to heat a certain amount of electric current. As may be imagined, it is much more difficult to prove that muscles produce electricity than to show that they give rise to heat, for to demonstrate their electric currents we have to use a very delicate instrument, far more complicated than the thermometer, called the galvanometer.

Tested by the galvanometer it is found that muscles give rise to no electric disturbance so long as they are at rest, but the moment they go into activity or contract they are responsible for a current which at once makes itself evident in the galvanometer. These currents are of very low voltage, but they continue to run as long as the muscle is in activity.

Now the heart is a muscle—a hollow, complicated muscle—and each time it beats or contracts it gives rise to a feeble electric current which is well demonstrated by means of a mirror galvanometer in a dark room. The mirror galvanometer was invented by Lord Kelvin to indicate the presence of electric currents in submarine cables, but it has been found well adapted to display currents of animal electricity. In this instrument a spot of light reflected from the mirror moves to the right hand or to the left on a white wall every time a current passes round the galvanometer.

Now since the heart is an automatically beating organ, and can continue to beat for many hours after it has been removed from a cold-blooded animal, the electric currents to which it gives rise will continue to swing the mirror so long as the heart itself survives. Indeed, the mirror continues to swing for some hours after all visible signs of life have left the heart. It is an impressive sight: in the stillness of the dark room the spot of light maintains its silent oscillations at the exact rhythm of the beating of the tiny living heart of the long-dead frog.

The beating heart of the embryo chick in the egg, when only just formed, can actually produce sufficient electric disturbance to influence the galvanometer.

If we bring the heart to a standstill, without killing it, by the application of some appropriate drug, then no electricity is manifested until the beating is resumed.

The latest development in the study of animal electricity is to photograph the currents of the beating human heart. This is done to-day by means of a very delicate and costly apparatus called the " string " galvanometer, because in it a fine filament oscillates. In the original method the oscillations of mercury in the capillary electrometer were photographed. The record is called an electrocardiogram.

Clinical medicine has thus been provided with a new method of diagnosis in heart disease, for the form of the record becomes noticeably altered in abnormal conditions of that organ.

One of the most interesting demonstrations of electricity of animal origin can be made with the retina of the living eye. Having killed a frog, we remove an eye and place it in connexion with the galvanometer in the dark room. The excised eye will live quite long enough to show a current when a lighted taper is held for an instant opposite the pupil. The retina in the depths of the eye responds to the stimulus of light by producing an electric current, and curiously enough yields an increase of current when the light is withdrawn.

All living tissues respond to stimulation by evolving electricity. For instance, when a nerve conducts its impulses, the only easily observable phenomenon is a series of electric currents in it. When a gland secretes, when a muscle twitches, when the heart beats, when the eye sees, electricity is always produced. As might be supposed, the absolute intensity of these currents is very small, of the order only of 0.01 volt. The late Dr Waller showed that so relatively inactive a tissue as the crystalline lens of the eye could nevertheless produce an electric current if the stimulus (an electric one) was sufficiently energetic.

There *is*, then, such a thing as animal electricity; but this electricity of animal origin is a physiological fact that has been overlaid by more nonsense than almost any other. The fantastic absurdity called "electro biology" was much heard of sixty or seventy years ago.

Animal tissues produce no "magnetism", as the physicists understand that term. The supposed power of magnets to induce the hypnotic state was long ago shown to be possessed by all sorts of agents that have nothing whatever magnetic about them.

All living things, vegetable as well as animal, can produce electricity; leaves, flowers, fruits if stimulated, can give rise to a current. It has actually been proved that a young unripe apple gives a better response than an old ripe one.

The man who discovered animal electricity was an Italian, Luigi Galvani, Professor of Anatomy in the University of Bologna. He discovered it accidentally. Galvani noticed that when the living hind legs of a dead frog, which were hanging over a copper hook laid on an iron railing, came in contact with the iron, the leg-muscles twitched. Galvani at once concluded that these convulsions of the legs were somehow an expression of animal electricity, but Alessandro Volta, his brother Professor in the University of Pavia, proved that, in the experiment just quoted, there is indeed electricity developed—not, however, by the living tissues, but at the contact of the two dissimilar metals, the copper hook and the iron railing. When the toes, blown against one of the uprights of the railing, completed the circuit through the legs, their muscles twitched.

Galvani nevertheless persevered in his efforts to demonstrate the reality of electricity of animal origin, and succeeded in devising a number of convincing experiments known as "contractions without metals". Quite the most striking of these is to allow the nerve supplying a muscle to lie over the isolated beating heart. At each beat of the heart, the current produced by the heart stimulates the nerve which in its turn causes the muscle to twitch: it is a very curious sight to see the living leg of a dead frog jerking at the exact rhythm of the frog's excised heart.

Curiously enough, it is in the group of the fishes that we find the most spectacular production of electricity. One of these, an eel-like creature, a native of certain warm South American rivers, can give a shock whose intensity is at least 400 volts. A specimen of one of these fishes was placed in the new aquarium at the Zoological Gardens, London. These electric eels are a family (Torpedo) which has developed this curious power as a means of protection from their enemies, or for the purpose of killing their prey.

The microscope makes it quite clear that in the electric fishes certain muscles have been profoundly altered so that, instead of producing movement and heat as ordinary muscles do, they manufacture electricity but at an intensity vastly greater than does any other living tissue.

Now just as ordinary muscle depends for its activity on impulses reaching it by means of motor nerves from nerve-centres, so too the electric organ is under the command of the cells of a nerve-centre. In fact, the nerve-cell in the spinal cord which " fires off " the electric battery of one side is by far the largest nerve-cell known. Each of these cells has one huge outgrowth of fibre which, proceeding down towards the battery, breaks up into a large number of small fibres for the various plates or discs of the electric organ.

The fish can discharge its salvos very much after the manner of a battleship discharging hers. It always discharges the shocks when it is handled or frightened, but it can also produce them by voluntary effort. The discharge is not of the nature of a constant current, it is rather that of a series of recurring impulses. They are quite powerful enough to give a man a most disagreeable shock. By means of suitable apparatus the currents can be made to ring a bell or cause a lamp to glow.

We are forced to admire the fertility of

Nature, how in one instance she can derive a maximum of heat and a minimum of electricity from a muscle, and in another, having meanwhile altered that muscle, can now obtain a minimum of heat and a maximum of electricity. Truly her ways are worth finding out!

CHAPTER III

ANIMAL HEAT

ONE of the most noticeable features of an animal is that it can produce heat. Biologists long ago came to the conclusion that it was by the oxidation, chiefly of the carbon and hydrogen in or by the living substance, that animal heat was generated. By "long ago" we mean the time of the great chemist Lavoisier (1780), who, though he made some mistakes in animal chemistry, certainly understood that animal heat and the heat of combustion were alike in this, that they were both due to oxidations. Before the discovery of oxygen by Priestley and Lavoisier, (1774), this position could not have been attained, although both Lower and Mayow of Oxford by 1669 had arrived at the conclusion that there was something in the atmosphere (Mayow called it "nitro-aerial particles") which was essential for the maintenance of animal life, and for all sorts of combustions. This was the dawn ; Priestley and Lavoisier were morning stars ; we live to-day in the high noon of the understanding of animal heat.

The chemical view of the origin of animal heat was, however, lost sight of by two such eminent thinkers of the eighteenth century as Boerhaave and Hales, who held that the heat of the body was due to the friction between the blood-corpuscles and the blood, and between them and the walls of the vessels. This is known as the iatro-physical theory of animal heat. One can do no more than mention the famous view of Stahl (1687), that the principle of fire was "phlogiston", a Teutonic will-o'-the-wisp that retarded the advance of thermo- and bio-chemistry for quite a hundred years.

But if by "long ago" we mean the centuries before the Christian era, we shall find that Aristotle, for instance, taught that animal heat was something absolutely unique. Aristotle believed that animal heat was something *sui generis*, something of an essentially different order from heat of non-vital origin—the heat of a fire for instance, or of a gas compressed, or of friction. He thought that the ultimate source of bodily heat was an eternal essence, the source of being of heaven and the stars, which latter were imagined as intelligent existences. This was, of course, the foundation of astrology, the notion that these celestial intelligences could interfere in human affairs, and were capable, through the harmony of their movements, of creating music, the " music of the spheres ".

The idea that prevailed for a very long time was that the blood, recuperated by the food in the intestines, was kept fluid in the heart by the heat there, but that the heart, becoming too hot, had to have cold air sucked in by the lungs to cool it. This teaching actually lingered until the time of Harvey, the discoverer of the circulation of the blood, who died in 1657.

Animal heat, as distinguished from every other kind, was thought to have quite special curative powers. Thus we are told in the Book of Kings that when King David " was old and stricken in years . . . they covered him with clothes but he gat no heat ". His servants recommended that a young girl should be brought to impart heat to him; " let her lie in thy bosom that my Lord the King may get heat ".

Our present view is that one of the abilities of living matter is to oxidize carbon and hydrogen to form carbon dioxide and water, which process cannot be effected without the evolution of a certain amount of heat. One of Lavoisier's ideas was that all the animal heat was produced in the blood of the lungs, that the lungs were the furnace of the body, and that the heat was carried thence by the circulating blood. We now know that the blood itself is responsible for the production of very little heat, and are quite certain that it is in the muscles that by far the greatest amount of bodily heat is manufactured. That a bloodless muscle can produce heat even after it is cut out of the body was proved by Helmholtz for the first time in 1848.

Everyone is familiar with the distinction between cold-blooded and warm-blooded animals; but this classification has lost its meaning. We now speak of animals which are able to maintain a constant temperature, and of those which are unable to do so, and these correspond to the old warm- and cold-blooded animals respectively. Those animals whose body temperature rises and falls as the temperature of their environment rises and falls are now known as poikilothermic, and those which are able to maintain a constant temperature, however much (within limits) the temperature of the environment rises or falls, are known as homoiothermic animals.

This classification has been found necessary because of facts such as the following: the temperature inside a hive of bees, insects classed as "cold-blooded", may even in winter rise to 21° C. above that of the outer air; a mammal, a "warm-blooded" animal, when hibernating, has a temperature almost the same as that of its cold environment; the temperature of a newly born mouse goes up and down with that of its surroundings; a frog or fish placed in warm water becomes as warm as the water. In other words a cold-blooded animal can, in certain circumstances, become warm-blooded, and a warm-blooded animal can become cold-blooded; so that all seems confusion.

The important point, of course, is not whether the animal's body feels hot or cold to human touch at any given moment of observation, but whether the animal can or cannot maintain a constant temperature while the temperature of its environment varies.

Healthy birds and mammals are able to keep their body temperature the same, no matter how high or low (within limits) the temperature of the environment goes. The temperature of all other creatures rises with a rise and falls with a fall of the outside temperature. In this latter class we must include the human infant, hibernating mammals, and newly hatched birds.

The power possessed by all living things, plants included, to produce heat as one of the manifestations of their vitality is referred to as thermogenesis. No doubt the actual amounts of heat produced by different animals are very different: a fish in unit time per unit of its weight is producing vastly less heat than a man, and a man less than a bird; but they are all producing heat.

The chief thermogenic sources are muscles (including the heart), the liver and the other glands; these, and not the blood, are the chief seats of heat-production. A frog whose blood has been replaced by a weak solution of common salt still continues to excrete carbon dioxide for some days, showing that oxidations are still going on in it although it has no blood at all. It is just as certain that an animal producing carbon dioxide is producing heat by the oxidation of carbon somewhere within its body as that a candle which is liberating carbon dioxide is producing its heat through the oxidation of the carbon of its fat.

Measurement of the quantity of heat is effected by means of the calorimeter, the principle being that the heat produced by any particular substance is absorbed by a known weight of water, which is, in consequence, raised a certain number of degrees in temperature.[1] A water-calorimeter is, then, essentially a metal box surrounded by a known quantity of water. To ascertain how much heat an animal gives out in a certain time, it is only necessary to place it in the inner box, which is, of course, properly ventilated. The heat that radiates from its skin is conducted through the box to the water; the heat of its expired air is absorbed by the water, the temperature of which, we shall say, has risen through n degrees. We shall suppose that the animal has done no "external work" while it has been in the calorimeter. If the weight of water is w units, then w multiplied by n is equal to the number of calories of heat lost by the resting animal in a given time.

[1] The large calorie, or unit of heat, is defined as the quantity of heat transferred to one kilogramme of water (2.2 lb.) in order to raise it one degree Centigrade in temperature.

The problem might now be faced, From what materials does the body manufacture heat? Since the body only very exceptionally absorbs any heat from outside sources—heat of the Tropics or a Turkish bath—it is quite clear that it must constantly be taking in the wherewithal to manufacture heat so continuously. Except in starving animals the source of heat is the oxidation of the food.

Just as a steam-engine transforms the potential chemical energy of the coal into the kinetic forms of energy—heat and external work (movement)—so the animal body transforms the potential energy of the food into the active forms of heat and muscular movement. Both are machines for transforming energy; the one is of iron, the other of protoplasm, but in both equally is the great law of the Conservation of Energy obeyed. This may be said to have been one of the mathematico-physical intuitions of that great mathematico-physical physiologist Helmholtz. For a long time it was thought that the animal organism was outside the scope of this great generalization; but it is known that an animal in a calorimeter will produce as many calories of heat in a given time as would have been generated by the burning of a weight of food equal to that of the food utilized during the period of observation. It is only a question of difference of rate in the setting free of the heat ; in burning the food we get all the heat out at once, in digesting it we get out just as much heat, but it is liberated much more slowly. The inhabitants of cold countries are pre-eminently fat-eaters ; the notion that fats give heat and sugars other forms of energy is probably not altogether fallacious.

The actual amount of heat that each adult loses per twenty-four hours is, in round numbers, about three thousand calories, the ultimate source of which is the oxidation of the various oxidizable chemical elements in the food of twenty-four hours. Knowing that one gramme of protein (flesh) yields us 4 calories, one gramme of sugar 4, and one gramme of fat 9, it is not difficult to construct a dietary containing, potentially, the necessary heat.

The portions of the total heat-loss borne by each system have been determined as follows—

Heat radiated and conducted from the skin, 2190 calories, or 73 per cent. of the total.

Heat lost in evaporating water from the skin, 435 calories, or 14.5 per cent.

Heat lost in evaporating water from the lungs, 216 calories, or 7.2 per cent.

Heat lost in warming the expired air to body-temperature, 105 calories, or 3.5 per cent.

Heat lost in warming the dejecta, 54 calories, or 1.8 per cent.

Of recent years a great many experimental studies have been made on the precise quantity of heat liberated by a healthy person completely at rest in bed (but not asleep) and taking no food. This amount of heat is usually stated in calories per square metre of the skin per hour. The method gives us the true heat-production of the tissues uncomplicated by the changing conditions, such as the taking of meals, the variations of temperature, muscular work and other factors. It is called the "basal metabolism"—a better name would be, the resting minimal heat production. Already it has provided data valuable in scientific medical diagnosis. In round numbers it is about 100 calories per hour.

In human beings the skin is *the* system of heat-loss, being responsible for the loss of 87.5 per cent. of all the heat lost. This is not so in some animals, in the dog, for instance, whose hairy coat does not permit its skin to perspire. The dog loses heat largely by its expired air, and by the radiation and evaporation of water from its tongue; thus on a hot day it pants and puts out its tongue.

Obviously our clothes prevent the loss of heat, and the more effectually the worse they are as conductors of heat. For this reason flannel, wool and furs are so much "warmer" than linen or cotton, which are materials from the vegetable kingdom, and therefore not the natural clothing of animals.

An animal is not only a transformer of energy, it is the most economical transformer known; for whereas in the very best steam-engines we obtain about 12 per cent. of the original potential energy as motion (external work), in the animal we get as much as 25 per cent. in the form of useful work. And whereas in the engine a great deal of the heat set free is lost or wasted from the engineer's point of view, the animal heat, so far from being wasted, is essential to the protoplasm to provide it with the optimum temperature for the performance of its vital activities.

The muscles as energy transformers differ from an engine in possessing two distinct powers, heat-producing and work-producing, or the thermogenic and dynamogenic powers respectively. Now the interesting thing is that, although a muscle cannot do work without producing heat, it can continue to produce heat without actively contracting, without doing work. In this latter state it is said to be in a condition of tonus or tone. No doubt when sufficiently analyzed, tonus is discovered to be a state of imperceptible or incipient contraction. Again, when a muscle having shortened continues to support a weight, but shortens no more, it continues to produce heat although in the sense of the physicists it is not doing any external work: thus dynamogenesis and thermogenesis are separate capabilities. But further, the two related capacities of muscle for heat-production and work-production only vary *pari passu* within certain limits.

As a muscle lifts heavier and heavier weights, it sets free more and more heat each time; and also if it raises the same weight each time, but under the influence of increasingly strong stimuli, it will evolve more and more heat each time. In this latter case the dynamogenic effects are the same throughout the series, while the thermogenic effects are greater and greater. As fatigue supervenes, the heat-producing faculty becomes impaired long before the work-producing. Thus very tired muscles may be able to lift the load to the height to which they raised it when fresh, but they now do so with the greatest possible economy as regards their store of potential energy, for they evolve the minimal quantity of heat. They oxidize material to as slight a degree as possible. Muscular work and heat are thus not quantitatively parallel products: there is nothing comparable with this in a machine of human construction.

As regards the kinds of food from which we derive heat, our ideas have undergone changes since the time when the German chemist Liebig divided foods into flesh-formers and heat-givers; the "meaty" stuffs and cheese he placed in the former group, the starches, sugars and fats in the latter. After having been much criticized, this classification by Liebig is admitted to be substantially correct. While sugars and fats do not build up tissue, but are found to be normally oxidized more or less directly for heat-giving or work-producing purposes, the proteins (flesh-formers) both repair tissue waste (by their nitrogen-containing moiety), and also contribute to heat-production by the non-nitrogen-containing substances into which in digestion they are split up. Liebig told the truth, but not the whole truth.

If during a certain time a person's temperature remains constant, that person is losing as much heat as he is producing. In technical language, thermogenesis, or heat-production, is just balanced by thermolysis, or heat-loss. Obviously, since we are constantly producing heat, unless we were as constantly to lose it, we should get hotter and hotter, and would damage our tissues by fatal fever. If no heat were lost at all, our temperature would rise 1° C. in half an hour ; in thirty-six hours our body fluids would be boiling ; in another thirty-six, autocremation would be far advanced. Now it is very well known that the temperature of a healthy man (98.4° F., 37° C.) hardly differs by a degree from one year's end to the other, or from the Poles to the Equator.

We may now ask ourselves the question, how it comes about that some animals have, and others have not, the power to keep their body temperature constant. Does not a fall of temperature necessarily depress vitality, and a rise, within limits, exalt it ? The answer is that protoplasm as protoplasm is certainly depressed by a fall and stimulated by a rise of temperature, which can be well seen by the slowing effect of chilling an isolated frog-heart and the accelerating effect of warming it. It would seem, then, there ought to be no such things as animals of constant temperature (warm-blooded). The solution to the puzzle is the high development of the nervous system in animals known as homoiothermic. In a mammal, when the temperature falls, the first result is loss of heat from the skin and an incipient depression of the temperature, but, through the nerves of the skin, impulses are sent into the nervous system which emerge as reflex stimuli to cause contraction of the skin vessels and increased heat-production in the muscles. The constriction of the skin vessels shuts some blood out of the skin and so diminishes heat loss, while the increased tone of the muscles increases heat-production, and thus the loss of heat due to the fall of temperature is so rapidly compensated that the temperature does not finally fall. Shivering is the familiar reflex effort on the part of the muscles to increase their heat as a result of the heat loss.

This power of balancing thermogenesis against thermolysis we call "thermotaxis"; the homoiothermic animals have, while the poikilothermic animals have not, the power of thermotaxis. Now, fever (pyrexia) is the upsetting of thermotaxis, the disturbance of this beautiful thermic balance. Theoretically, fever, or a rise of temperature, may occur if the heat production is too rapid for the heat loss, or, on the other hand, the heat production being unaltered, if the heat loss is diminished in rate. Both these states of the upset thermal balance occur. Sir William Hale-White has shown that whereas in pneumonia and erysipelas the fever is due to increased heat-production, in typhoid fever and in suppuration the rise of temperature is due to a diminution in the heat loss.

Fever is to-day regarded by physicians in a totally different light from what it was even a few years ago—in itself a wholly bad thing, to be reduced at any cost. The increased heat production is looked on as a reaction on the part of the living cells to the noxious stimulus of the micro-organisms or their soluble poisons, a response of a protective nature. Hence the indiscriminate lowering of the temperature by drugs (antipyretics) is not now nearly so common as it used to be. It is recognized as possible that the increase of heat (fever) may be evidence of sufficient vitality on the part of the living protoplasm to withstand the assaults of the infective agents, the increased heat being the biophysical response to the micro-organic insults. The drugs which benefit fever most are now regarded as doing so, not because they lower the temperature, but because they attack the specific cause of the malady; quinine, for instance, in malaria destroying the parasitic *Plasmodium malariæ*, the salicylates antagonizing the poison of rheumatic fever, and so forth. Those versed in vegetable physiology have been able to show that even in the case of parasites attacking trees there is a rise of temperature as a reaction to these assaults, and therefore botanists actually speak of " fever " in plants.

Of course, it is not to be imagined that in no circumstances is fever, or very hot blood, injurious to the body. Within the last few years definite experiments have been made showing that blood hotter than a certain temperature does permanently damage the cells of the central nervous system.

A short exposure to 47° C. or a longer one to 42° C. kills the cells of an animal's brain by coagulating one of the essential constituents of their living substance. "Sunstroke", as it is called, is the result of the too hot blood injuring the cells of the brain, especially those related to consciousness. When the cells are only slightly injured the person may recover, and be "a little queer in the head" for the rest of his life; if, however, the cells are decidedly overheated, as in "heat-stroke" or "heat-apoplexy", death in collapse supervenes, the person never regaining consciousness. Heat-stroke of this kind may occur in places to which the sun never gains access, as, for instance, in front of the furnaces of a steamer in the Red Sea.

On the other hand, depression of the temperature of the blood below its normal is as fatal, although not so rapidly, and for quite other reasons. Great loss of heat depresses the tissues so that death results. What is known to coroners' juries as, "death from exposure" is really due to heat-loss. An underfed, poorly clothed, and perhaps also alcohol-intoxicated person falls asleep out of doors on a frosty night; so much heat is lost that the heart and nervous system never recover; the person never wakes. The lowest (subnormal) temperature in man—namely, 80° F.—that has been reached, has been recorded under these conditions.

CHAPTER IV

PRECONCEIVED PERFECTION

ATTEMPTS have been made to hinder the progress of science by many things but by none more than by preconceptions. These ideas were very often assertions purporting to embody some religious opinion or doctrine to which everything new must conform or forthwith be rejected.

For instance, it was assumed by the early Greek philosophers that, since the circle was a "perfect" figure and the heavenly bodies were also perfect—as seemed then self-evident—these heavenly bodies must revolve in circles. The Ptolemaic system in astronomy was based on this assumption. There was thus considerable opposition to the demonstration that the planets revolved in ellipses. These perfect bodies moving in perfect circles were supposed to make perfect harmonies—"the music of the spheres". No one had heard it; but that did not matter; it must be perfect music since it was produced in heavenly places.

Galileo was greatly worried over these preconceived notions about perfection. The sun was perfect—that was self-evident ; yet Galileo had been rash and irreverent enough to discover " spots " on the luminary. These spots were physical appearances discernible by the telescope; but Galileo's critics, choosing to take the term in the sense of moral blemishes, condemned the astronomer as a thoroughly impious person.

This ascription of moral worth to inert objects has been the cause of a good deal of trouble from time to time in the history of progress.

In the case of Galileo, it was not simply that he asserted the earth moved round the sun, but that he had declared the heavenly bodies, like this terrestrial globe, "corruptible ".

He might have been warned by the fate of Copernicus, whose work, disproving that this small planet was the centre of the universe, was placed on *The Index*. The theologians seem to have been very sensitive about astronomical heresy. Bruno was burnt because he had speculated, amongst other things, on the " plurality of worlds ".

The orthodox view was geocentric and anthropocentric: this planet was the most important in creation, and man the most important being in it.

This sort of thing we find also in Aristotle's view of animal heat. Aristotle regarded the heat of the animal, and particularly the heat of the human body, as of a far finer essence than heat of non-vital origin. The heat in the human body was supposed to be something allied in its essence to the energy displayed by the stars, which, regarded as sentient beings, were held to preside over human destiny. The expressions, "ill-starred" and "my stars!" refer to this belief.

The view lingered for a long time, although the acute mind of Shakespeare evidently doubted it, for he makes Cassius say—

> The fault, dear Brutus, is not in our stars
> But in ourselves that we are underlings.

Notions of preconceived perfection troubled the great William Harvey, the discoverer of the circulation of the blood. It is, of course, of the essence of the Harveian doctrine that the blood, the same blood, which leaves the left side of the heart by the arteries for distribution to the body, returns by the veins to the right side of the heart. It is like the procession of stage-warriors: there may be only a dozen of them; but as they go out at one door and come in by another they create the impression of a never-ending procession; yet it is always the same twelve men.

One of Harvey's correspondents, Dr Caspar Hoffmann of Nüremberg, had objected to the idea of the circulation, because he considered that Harvey had

> impeached Nature of folly and error, had imputed to her the character of a most clumsy and inefficient artificer in suffering the blood to become recrudescent and making it return again and again to the heart in order to be reconcocted, to grow effete, as often, in the general system, thus uselessly spoiling the perfectly made blood merely to find her something to do.

Hoffmann's amazing criticism amounts to this: The blood cannot keep on returning to the heart because that indicates an imperfect arrangement; now the arrangements of the heart are perfect; therefore there is no circulation. The results of experimental research could not be correct because, forsooth, Hoffmann believed that the heart works "perfectly" in some manner other than that involving a circulation of blood through it. This is the *à priori* argument *in excelsis*. It is extremely typical of the sort of reasoning that proceeds from preconceived ideas. It was employed to quite within our own day.

Another excellent example is the criticism which Stensen had to endure for saying that the heart was a muscle. Nicholas Stensen, the Dane, was a doctor of medicine and an eminent man of science who studied anatomy and biology experimentally with great success. His dissections showed him that the heart was a complicated arrangement of muscular fibres; that it was, in fact, a hollow muscle. His critics, however, held that as the soul was in the heart, it was impossible that the heart could be a "common muscle"; the mere suggestion was rank impiety. Stensen was duly censured for such irreverence. To us to-day it seems ludicrous that such puerile criticism should have stood in the way of physiological discovery; but the strangling power of preconceived ideas is vast.

The progress of anatomy in the Middle Ages was very much impeded by the ban placed upon dissection by the Mohammedan injunction against it. The Koran decreed that to touch a dead body made a person ceremonially unclean; dissection, especially of the human body, was consequently impossible. The Jews had no greater liberty in this respect than the Arabs. The result was that the Jewish and Arabian doctors of the Middle Ages kept on repeating the statements of the Greek and Roman writers on anatomy. They verified nothing and they discovered nothing.

It was one of the minor results of the Renaissance that the study of *practical* anatomy was revived and introduced first into the medical schools of the Continent and then after a due interval into England.

The case of Tagliacozzi is typical of opposition to science through preconceived notions about theoretical perfection. Gasparo Tagliacozzi was an Italian surgeon who flourished in the middle of the sixteenth century, and who had specialized in those operations known as "plastic", whereby he repaired noses, ears and lips by grafting skin from the arm. The operation for a new nose (rhinoplasty) was his most famous achievement. He had plenty of opportunity to perform such operations, for both in duelling and in the perpetual wars of that period, the most prominent feature of the face was apt to be mutilated.

By his scientific brethren his work was thought very highly of, and he is commemorated by a life-sized statue in the anatomical theatre of the University of Bologna. But in other circles it aroused the fiercest opposition: he was denounced for his impiety in daring to alter the human countenance "made in the image of God"! The excitement even outlasted his life; for after many years in the grave, his bones were dug up and scattered by order of the ecclesiastical authorities. The gravamen of the charge against Tagliacozzi was his irreverence in interfering with the human face; it would seem to have been quite allowable to mutilate this sacred part of the body in war, but thoroughly immoral to try to remedy the damage.

Benjamin Franklin demonstrated the identity of lightning and electricity, and recommended the use of lightning-conductors; but he had the greatest possible difficulty in overcoming resistance to his recommendations on the ground that he was interfering with the divinely sent thunder and lightning.

Yet another case of the same sort of thing

is the opposition to Jenner's discovery of the efficacy of vaccination against small-pox. Jenner had many trials to bear; in some quarters it was not thanks he got for his trouble in showing mankind how they might avoid one of the most dreadful scourges that have ever afflicted it.

A Rev. Dr Rowley, who otherwise would have remained unheard of, evidently imagining himself honoured by special participation in the divine counsels, declared—

> Small-pox is a visitation from God, but cow-pox is produced by presumptuous man. The former is what Heaven has ordained, the latter is a daring violation of our holy religion.

Such nonsense is intolerable. How could Jenner possibly be responsible for cow-pox, which, as a disease of cattle, was in existence æons before he was born?

To concoct theoretical, *à priori* statements about the supposed divine or other origin of disease was nothing short of an outrage on an investigator who was endeavouring to discover a remedy for a world-wide scourge that had preyed upon mankind from the very dawn of history.

But before the time of Jenner there was similar trouble over inoculation of small-pox itself. This practice was based on the fact, familiar enough to everyone in the reign of George I, that anyone who had had small-pox was extremely unlikely to get it a second time. The Turks, Circassians and Chinese had for long inoculated small-pox in the hope that by having it mildly they would be immune from it when an epidemic came.

Lady Mary Wortley Montague, wife of the British Ambassador at Constantinople, had been so impressed with the immunity from serious small-pox conferred by the previous inoculation of it, that she caused her infant, born in Constantinople, to be inoculated there.

But, as Voltaire, in his "Letters on England," says—

> The chaplain represented to this lady, but to no purpose, that this was an unchristian operation, and therefore that it could succeed only with none but infidels.

Truly a discriminating specificity of action —a new test for believers! As is well known, Lady Mary introduced inoculation into England and thereby undoubtedly saved thousands of lives.

The story of the opposition to the views of Franz Joseph Gall, the founder of the pseudo-science of phrenology, is another case in point. Gall was not objected to because he taught an absurd system of cerebral physiology, but because he had outraged religion in Vienna by teaching that mental and moral qualities had their physical basis in the brain. He was "materialistic" because he asserted that mental and moral states were conditioned by (brain) matter; and to be "materialistic" in 1800 was the next thing to being an atheist. It was well for Oliver Wendell Holmes, when he published his *Mechanism in Thought and Morals*, that he did not live in Vienna in 1800.

Gall was groping after what we now call the localization of functions in the brain; and he was the first to hint that there was a centre for speech in the grey matter, a region of the organ of the mind devoted to the expression of ideas by appropriate words. All subsequent work has confirmed his contention. The title of his objectionable work was: "The Anatomy and Physiology of the Nervous System in General and of the Brain in Particular, with Observations on the Possibility of Recognizing Various Intellectual and Moral Dispositions of Man and of Animals by the Configuration of their Heads ".

To attribute man's moral qualities to any physical organ was reprehensible in a high degree; and so Gall was driven from righteously scandalized Austria and took refuge in Paris, where he practised as a physician and lectured on the brain until his death in 1828.

Had the experimental physiology of the brain and nervous system been in a more advanced state when Gall was studying cerebral localization, the chances are he would not have lost his way as he did in the intellectual morass of phrenology.

One of the best examples of attempts to obstruct the beneficent progess of medical science, by alleging it contrary to nature and to religion, is that of the introduction of chloroform.

It was assumed that pain must not be removed by any " artificial " means, that to abolish pain even of childbirth was " con trary to Nature ". Just as Tagliacozzi had been denounced as sacrilegious for tampering remedially with the human countenance, so Simspon was villified for endeavouring to mitigate human pain. When a lady told him that it was contrary to Nature for him to remove the pains of childbirth, he replied—

> It was 'contrary to Nature' for you to come over the water in a steamer to see me.

When the Scottish clergy quoted from Genesis the curse on Eve—"in sorrow shalt thou bring forth",—Simpson quoted against them: "And the Lord God caused a deep sleep to fall upon Adam, and he took one of his ribs and closed up the flesh instead thereof". From this Simpson argued that prior to this very early case of resection of a rib there was due preparation made for it by general anæsthesia.

If we did nothing except what is conformable to a state of Nature, we should still be wrapped in sheepskins and be living in the dens and caves of the earth.

CHAPTER V

CIRCULATIONS

A CIRCULATION is the return of a moving object to the place where it started.

When in the theatre they wish to represent the marching of an army, and they have only four-and-twenty stage warriors (who cannot be expected to be " terrible as an army with banners ") the stage-manager makes these four-and-twenty men enter the stage by one door and leave it by another. If they do this sufficiently often, it gives a spectator the impression of the marching of a large body of men: the illusion of quantity was created by a circulation.

Bank-notes circulate, because after no matter how many transferences from one person to another, the notes return to the bank which issued them.

The circulation of water is one of the most obvious of Nature's great cosmic circulations.

We know that the water of the sea, lakes,

rivers and of the damp earth is being constantly evaporated into the atmosphere where it may remain unseen for a long time.

Sooner or later it is condensed into some visible form, dew, mist, fog or cloud.

The particles of the clouds run together, and, becoming too heavy to float, fall to the earth as rain, snow, sleet or hail.

It is certain that the *same* particles of water have been doing this for millions of years.

It is quite possible that a particle of the water of a cloud may be one of many million particles drunk by an animal on some occasion; it then passed into the animal's blood and was carried through that long circuit of the arteries, capillaries and veins of the body; finally it reached the kidney whence it was thrown out of the body, or reached the lung, whence it was exhaled as a particle of vapour such as is seen on a cold or very damp day.

The biblical description of "the rain that cometh down and the snow from heaven, and returneth not thither" is true only of one aspect of the circulation of water.

The circulation of the blood is, of course, the most familiar example of matter returning to the place whence it started. A particle of blood leaves the heart, passes by the arteries to the capillaries, thence to the veins and so back to the heart again—and this ceaselessly.

It is the *same* blood.

Harvey insisted that if it is *not* the same blood which keeps on going round and round, we are landed in the absurdity of having to admit that in (say) an hour, the heart will pump out into the body very much more blood than the body of the animal is known to possess. This consideration he gave as one of the proofs of the circulation.

The body of a man has about three and a half litres of blood, let us say five at the most; the heart throws out at each beat about 60 cubic centimetres; now there are about 70 beats per minute, so that the heart per hour deals with 60 by 70 by 60 or 252,000 cubic centimetres, or 2,520 litres. But the whole body contains certainly not more than five litres; so that unless the same blood returns to the heart, the heart in an hour would eject about 500 times as much blood as the body possesses, "which is absurd!"

Take now the case of an atom of carbon in a piece of cane-sugar which we have just eaten. It is digested, and passing into the blood goes to the liver, where it may be stored for days in the form of a kind of starch. *Finally it leaves the liver as sugar again*, and once more entering the blood-stream is carried to the muscles and other tissues, where it is oxidized to carbon dioxide. This gas enters the blood, and being carried to the lungs is there exhaled into the atmosphere as an invisible molecule of gas. The same carbon atom therefore which began its journey in a solid piece of sugar is now floating in the air as part of a molecule of gas. How long it may float about no one can tell; but at last it will be taken in by some green plant, and under the influence of sunlight on the living substance in the leaves will be united to hydrogen and oxygen and so condensed to form starch. This starch the plant finally turns into sugar, and so we are where we began.

This sort of thing has been going on ever since plants and animals lived together on the earth. A rather longer cycle is gone through by an atom of nitrogen which is one of the elements in the protein of the grass. The grass is eaten by a cow, and, we shall suppose, the beef becomes part of a human being. Some of the nitrogen of the human being is thrown out of the body as urea, a substance which in the soil is decomposed into ammonia and water. The roots of the grass absorb the ammonia; and its nitrogen is built up into the protein with which we started. This circulation of nitrogen can go on for ever.

But it is clear there is a circulation of matter all the time between plant and soil and soil and plant. This autumn's dead leaves decomposing at the foot of the oak and being absorbed by its roots, will become next summer's leaves and acorns. The body we bury to-day will sooner or later be broken up into its component salts, which, absorbed from "mother earth", will be reincorporated into future living beings. And what happens once can go on happening endlessly.

"Dust to dust and ashes to ashes" is the poetical form in which this is expressed in the Prayer-book. If translated into the language of modern science it expresses one of the most important truths concerning the intimate relations between three things, the soil, the living, and the dead.

"Imperial Caesar dead" might in time have become a constituent of the body of a slave!

CHAPTER VI

THE EYE AS A CAMERA

THERE are several very close resemblances between the mammalian eye and a photographic camera.

The essential parts of a camera are, a sensitive plate to receive the image of some object being photographed, and a lens to form that image. The camera, like all other optical instruments, is blackened inside.

If for the sensitive plate we substitute one of ground glass, then on that plate we can observe from behind the camera the small, inverted image of the person about to be photographed.

This is what the photographer sees when he throws a black cloth over his head and seems for a little while to be lost to the world at the back of the camera. He is looking at the minute image of the " sitter " shining through the translucent glass plate.

The double convex lens which is responsible for the image is possessed of such properties that if it be placed in front of a large, distant object, it will produce on a screen close to it a small, inverted image of that object: like Luther, "it cannot do otherwise".

Now the human eye is exactly such a camera: it has a double convex lens in front, and a sensitive plate (the retina) behind, and it is blackened inside. The "black eye" of popular acception refers not to this internal pigmentation but to an accidental, external disfigurement. The objects of the outer world are relatively far from the lens compared with the retina which is quite close to the lens.

It follows from this state of matters that the image of the external world cast by the lens upon the retina must be very small, and also completely upside down. This fact can be proved if one does not grudge a rather troublesome dissection as follows.

If we fix the eye of a sheep or ox a few feet away from a bright lamp, and then take a razor and shave away carefully the tough coat at the back of the eye (sclerotic) until the delicate retina comes into view, we can just make out the tiny image of the lamp upside down. We are looking at the retinal image shining through the dead retina just as the photographer sees the sitter's image shining through the dull glass plate which he uses before he inserts the sensitive plate itself. Though difficult to carry out, this observation is very instructive since it gives us ocular demonstration that there is in the eye a real image of the outside world and that it is, as regards that world, upside down. Yet we do not see the world upside down; it is only our accomplished writers of paradox who can turn the world upside down.

This is scarcely the place to discuss the problem why we do not see the outer world inverted. Both the psychologists and the metaphysicians have something to say about it, the gist of their remarks being that, as we have no direct knowledge of the existence of the retinal image at all, we cannot consciously be dealing with it. Unless it were for anatomists and physiologists, we should never know that there was such a thing as a retina or that it had an inverted image focused upon it.

The extreme minuteness of the image in

the eye is a fact which very few people realize. The entire surface of the retina is only about one square inch, and not by any means all of this surface receives the image of the outer world. The image is received on a specialized, slightly hollowed spot which is about 1/12th of an inch in diameter (the macula lutea). This really means that the image, say, of St Paul's Cathedral, viewed a few hundred yards away, is all condensed upon a tiny spot at the back of the eye about 0.0045 of a square inch, and that in this image the dome is down below and the steps and the pavement are up above.

It is well known that no lens can give a clear image of a near object and of a distant object at one and the same time. When a photographer wishes to focus a near object, he pushes the lens farther in, and when he wishes to focus a far object, he pulls the lens out. In the mammalian eye this bodily movement of the lens cannot take place on account of the way in which the lens is suspended in the eye.

But a very remarkable thing happens; the lens itself becomes thicker in the centre, that is a stronger lens when a near object has to be focused, and flatter in the centre, that is weaker, when a distant object has to be focused. All this, which happens quite outside our consciousness, has nothing analagous in the case of the camera ; in a certain sense it shows the very great superiority of the eye to the camera.

There is one more resemblance between the eye and the camera with which we may close this study. The camera has a number of metal discs called " stops " with circular holes in them. These diaphragms the photographer uses to sharpen the image. It is another property of a double convex lens that the rays passing through the margins of it tend to blur the image. When the photographer, therefore, wants to cut off these rays, he slips one of the stops in front of the lens so that the central rays pass through the aperture and form the image while the marginal rays are cut out and so prevented from taking part in the formation of the image.

Now the eye has an exactly similar apparatus in the circular iris with its central circular aperture the pupil. The opaque iris cuts off the marginal rays which are not wanted, while the pupil allows the image-forming rays to pass through the centre of the lens. In this way the iris sharpens the image exactly as does the photographer's "stop".

The size of the pupil is altered by the activity of certain microscopically small muscles in the iris; when the light is very bright the pupil becomes small, when the light is feeble it becomes dilated. This action goes on entirely outside our knowledge. The mechanism of the symmetrical closing in of the circular aperture in the iris is very perfect and has been imitated in the "iris-diaphragm" of the modern microscope.

Finally, why is the inside of the eye blackened? For the same reason that all optical instruments are blackened inside—to prevent internal reflexions, and the general glare that would result from having the interior of the eye incapable of absorbing stray light, that is, light not employed in forming the image.

CHAPTER VII

PREDICTION IN BIOLOGY

PROBABLY nothing impresses us with the dignity and soundness of science so much as the accurate fulfilling of some one of its definite predictions. In a mathematical science, such as astronomy, prediction may be said to be an inevitable part of the investigator's duties, since the phenomena he deals with are necessarily of a recurring order, but in any science it is a valuable method of putting a theory to the test or of estimating the value of a working hypothesis. For instance, the Russian chemist, Mendelieff, having arranged the elements known to chemists in certain natural series, which nevertheless contained gaps, predicted that before long these gaps would be filled up by elements yet to be discovered. Those of us who have followed the development of chemistry know that these chemical prophecies have come true. The objective reality of those properties on the possession of which Mendelieff's series was constructed, was thus established.

In the non-mathematical sciences, prediction may be said to be supplanted by prescience or prophecy. Thus, when the great William Harvey wrote to his friend Horst (1654), " Many things still lie hidden . . . destined to be drawn up into the light by the indefatigable diligence of coming ages ", he had some, but no adequate, idea of how truly he spoke as a prophet. In a vastly fuller way than he could possibly have imagined, this prescient utterance has been fulfilled. By the " indefatigable diligence " of the great Italian biologist, Marcello Malpighi, Harvey's own generalization as to the circulation of the blood was verified to ocular demonstration by means of the microscope (1660). In a thousand ways have Harvey's words been justified; has not " diligence " triumphed over all the technical difficulties in the way of preserving and analyzing the living tissues? Do we not by histo-chemical diligence preserve for ever from decay those very invisible blood-vessels which Harvey's eyes never saw, but which he was so certain existed? How transported with joy would he have been to know that we can fill them with a substance which, solidifying in them, demonstrates their presence with the utmost clearness. Diligence has literally brought them to light; the master's prophetic vision saw that which was invisible.

Those engaged in studying the intricacies of the nervous system seem at all times to have indulged in prediction. Thus, the Father of Anatomy, Vesalius himself, ventured to assert (1543) that by mere dissection of the brain no further discoveries as to its functions would be made. Three hundred years of anatomy have justified his belief in the comparative sterility of the study of the gross anatomy of the nervous system, at least as regards the discovery of function. Not by dissection, as Vesalius knew it, but by microscopical investigation, by inference from careful experiment and by injuries accidentally produced, have the activities of the brain been made clear to us. Vesalius has been justified in his prophecy that dissection is a barren art, but not justified if he meant to predict that the cerebral functions would be for ever beyond the ken of the students of living things.

A most prescient physiologist, the Dane, Nicholas Stensen, prophesied that, as regards the central nervous system, no further advances in understanding it would be made "without a special manner of preparing it". This was uttered in 1668, and has been completely justified. The preparation of the substance of the brain and spinal cord to fit them for microscopical examination was surrounded by so many difficulties that not for nearly two hundred years was it possible to make these structures sufficiently transparent. The difficulties in dealing with the cerebral substance were such that they overcame that Alexander of Biology, Malpighi, who declared that the brain was a gland which secreted spirits. So far astray is it possible to be led when the time has not yet come for the fulfilment of a prophecy!

Yet another seventeenth century prophet, but this time a false one! The great Sydenham, the English Hippocrates, though a contemporary of Harvey, was not a disciple of Harvey. He saw no use in Harvey's discovery, and consequently ignored it: in particular, he despised the study of the minute structure of tissues, or "histology" as it is called to-day.

Again, according to his scholarly biographer, the late Dr Payne, Sydenham prophesied " that the human intellect will never be able to understand the use of the different parts of the brain ". Supremely difficult as we know this study to be, surprising advances have been made in it, but just by the experimental method of Harvey rather than by the mere observational method of Sydenham. By that very method of research, the microscopical, which Sydenham trusted least, a very great deal of excellent physiological work has been achieved.

A contemporary of William Harvey and of Thomas Sydenham was a certain Gideon Harvey, " Physician to the Tower ", chiefly remembered for his coarse but amusing attacks on the College of Physicians. In what follows he describes the type of doctor of the school of his great namesake, a description which we can imagine Sydenham fully endorsing—

" They flay dogs and cats ; take livers, lungs, calves' brains, or other entrails, dry, roast, parboil them, steep them in vinegar, etc., and afterwards gaze on little particles of them through a microscope ; then obtrude to the world in print whatever false appearances gleamed into their eyes ; and all this to no other end than to beget a belief in people that they who have so profoundly dived into the bottomless pores of the parts must undeniably be skilled in curing their distempers."—("The Conclave of Physicians." Second edition, 1686.)

In the same strain of unbelief as to the possibility of histological demonstration wrote no less a man than the philosopher, John Locke, a close friend of Sydenham. He committed himself to this: "That Nature performs all her operations in the body by parts so minute and insensible that I think nobody will ever hope or pretend, even by the assistance of glasses or other inventions, to come to a sight of them".

Professor Osler, in his delightful essay on Locke, tells us a significant fact apropos of this remark: that on the top of the page on which it occurs there is a sentence in Sydenham's handwriting fully confirming his friend's scepticism. In modern language, Locke said there was no future for histology or bio-chemistry, which were futile attempts to unravel the intricacies of texture and of function wholly inaccessible to man. Now it is precisely in these two realms that the greatest conquests in medicine have been made, for the microscope and the test-tube have been two of the most powerful weapons in her peaceful armamentarium.

Sydenham was a great physician, a great observer in the realm of disease, but he was no prophet either in his own or any other country. He fell into an error, not uncommon in robust but not broad minds, of imagining that there cannot be anything of much value outside that which these minds readily understand.

Prophecy has always occupied an important place in scientific discovery.

Jenner prophesied that the patient whom he had previously vaccinated against small-pox on being inoculated with small-pox would fail to take it. This was the boy Phipps who had been inoculated by Jenner with the virus of cowpox (vaccinia) on 14th May, 1796, and on the following 1st of July had been inoculated with fluid from a vesicle of a case of genuine well-developed human small-pox: the boy did not take the disease at all. Jenner waited until the 19th of July, and, finding that the boy was still quite well, could no longer delay, but wrote to his friend Gardner as follows: "Listen to the most delightful part of my story. The boy has been inoculated for the small-pox which, as I *ventured to predict*,[1] produced no effect.

[1] The italics are mine.—D. F. F.-H.

I shall now pursue my experiments with redoubled ardour ".

One of the most interesting writers on the mind and its organs—the brain and the nerves—Prochaska, of Vienna, makes a remark of great interest with regard to those workers who were to come after him. It concerns the two roots of the spinal nerves. The structural differences between the two roots of a spinal nerve, the upper with its swelling or ganglion, the lower without any ganglion, had long been familiar to anatomists. Prochaska knew that one of the chief nerves of the head, the fifth or trifacial, had also two roots, one large and ganglionated, the other small and non-ganglionated. He was thus led to speak prophetically in the following words—

> " He who shall unravel the uses of the ganglia will also give a reason why the fifth pair of cerebral nerves pass through the semi-lunar ganglion (with the exception of the fasciculus which joins the third division) without touching the ganglia, and why only the posterior roots of the spinal nerves enter the ganglia, whilst the anterior roots pass by without any communication with them." (1784.)

Now this was precisely what Sir Charles Bell did twenty-seven years after the above was written. Bell in the first place showed that the upper or posterior roots of the spinal nerves were functionally exactly the opposite of what Prochaska supposed them to be, namely, nothing less than nerves whose ganglia prevented them conducting impulses into the nervous system. By conclusive experiments Bell proved that these posterior roots were, in point of fact, the only avenues *into* the spinal cord, and that their ganglia in no way interrupted impulses passing through them.

Of the nature of the nerve-impulse or *vis nervosa* as he called it, Prochaska wrote in these words: "Possibly the diligence of the very sagacious observers of Nature may discover whether that principle be electricity or phlogiston or some species of air or the matter of light or something compounded of these" (1784). Many "sagacious observers" since 1784 have done their best to discover the nature of the nervous impulse; and while physiologists would agree that it was not phlogiston or a species of air or the matter of light, not all would so readily put the first possibility out of court. With the formulation of the doctrines of electro-chemistry has risen an attempt to explain nervous activity, not indeed in terms of electricity as we know it flowing in wires, but in terms of electrically-charged particles—ions—constrained to set themselves in certain positions at certain moments in such a manner that electrical phenomena can be manifested.

A particularly interesting example of unfulfilled prophecy in biology occurs in the writings of no less a man than Johannes Müller, probably the most encyclopædic of Germany's scientific encyclopædists. He has been treating the nature of nervous action, and then he utters the following prediction as regards the velocity of propagation of the nerve-impulse: "We shall probably never attain the power of measuring the velocity of nervous action, for we have not the opportunity of comparing its propagation through immense spaces as we have in the case of light" (1835).

Müller believed, as did other leading biologists as late as 1835, that the velocity of the nervous impulse was something prodigiously great, of the same order of magnitude as that of light; indeed, that the act of thinking was performed with "the speed of light". We still speak of a thought "flashing into the mind", and of things occurring with "the speed of thought". Exactly fifteen years after this prophecy was made, Hermann Helmholtz, in 1850, measured the rapidity of the nerve-impulse in the motor nerves of the frog, and showed it to be perfectly calculable, not much more than about 90 feet a second. He subsequently measured it in human nerves, and found it to be about 360 feet a second. Once again we see the danger of venturing to predict any limitations to the possibilities of experimental science.

The last case of prophecy regarding the functions of the nervous system which we shall notice is one by the amateur physiologist George Henry Lewes, well known as the biographer of Goethe and the first husband of "George Eliot". He ventured to predict that when the posterior roots of the spinal nerves were cut, the erection of the hairs on an animal's back would be no longer possible, in other words, he committed himself to the belief on *à priori* ground that the posterior roots would be found to carry impulses *out* of the spinal cord.

The passages relating to this prophecy are so interesting in more than one respect, that it will be well to have them before us—

"The posterior nerves being distributed to the skin can, it is obvious, only exercise a motor function in as far as they are related to moving organs or muscles. Now in the skin there are such organs, insignificant enough and only discovered within the last few years; these are muscular fibres surrounding the hair-follicles, and it is these which the posterior nerves cause to contract. If there were no muscular fibres in the skin, the posterior nerves would have no motor function; but they would have the possibility of exercising such a function—the same in kind as that of the anterior nerves. Thus does anatomy assure us that both nerves are sensory and both nerves motor. With this torch in our hands let us enter the otherwise obscure path of experimental inquiry. . . . The conclusion seems irresistible that the muscular sensibility is derived through the anterior or muscle-nerves. . . . There has been no collection of facts to throw light on the movements of the muscular fibres in the skin, but it is worth inquiring what share they have in the erection of 'quills upon the fretful porcupine' or the hairs on the back of an irritated dog. Division of the posterior roots would, I should anticipate, prevent the quills of the porcupine from rising up. But there is no experimental evidence on this subject present. . . . There is no fundamental distinction between the two nerves, both are sensory and both are motor; but they are so in different degrees."

When the late Professor Langley and Sir Charles Sherrington in 1891 came to examine these roots experimentally, they found the exact opposite of this prophecy of Lewes to be the true state of matters. The "pilo-motor" fibres, as they called them, did *not* pass through the posterior roots, but left the spinal cord by the anterior, so that section of the posterior roots does *not* paralyze the hair-muscles. In fact, there is no real exception to the Bell-Magendie law that all impulses entering the cord do so by the posterior roots, and that impulses leaving it do so by the anterior roots.

The passages just quoted are an excellent example of library or *à priori* physiology. Anterior to all investigation by experiment, the author gets hold of a principle of functional symmetry which he thinks ought to hold good in the case of the two nerve-roots. It is a torch to light up the otherwise obscure path of experimental inquiry! Experimental inquiry, on being made, discovers the exact opposite of what the anatomical light indicated. Therefore, anatomical torches by themselves are not to be trusted; and further, however respectfully *à priori* philosophy should be treated, *à priori* physiology most emphatically is not to be encouraged.

Another definite prediction was Pasteur's regarding the prophylactic inoculations for anthrax in sheep and cattle in 1882. The incident is so well known that it need not be repeated in all its detail. Pasteur had just perfected his method of protecting animals from a virulent disease by repeated injections of the attenuated poison of the same disease. The experimental conditions, prescribed for him by an agricultural society, were that twenty-five sheep were to be inoculated against anthrax, and, after an interval, to receive a virulent culture of the germs of that disease; at the same time twenty-five normal (uninoculated) sheep were to receive some of the same virulent culture. Pasteur accepted the challenge with equanimity. "The twenty-five unvaccinated sheep will all perish", he predicted, "the twenty-five vaccinated ones will survive"—and lo, it was so. The crowds who came to Pouilly-le-Fort to see the crucial experiment fail and to jeer at the man of science were converted then and on that very spot; and "the whole of France", wrote Pasteur's son-in-law and biographer, Valery-Radot, "burst out in an explosion of enthusiasm".

CHAPTER VIII

THE SIGNIFICANCE OF "PLAGUE, PESTILENCE, FAMINE, AND SUDDEN DEATH" IN THE LITANY

CERTAIN parts of the Litany must often have struck us in church as peculiarly inapplicable to our present conditions.

This would apply with especial emphasis to the prayer, "From lightning and tempest; from plague, pestilence, and famine; from battle and murder, and from sudden death, Good Lord deliver us". But when we remember that these words were put together for the first Prayer Book of King Edward the Sixth, published in March, 1549, our opinion about their inapplicability should be considerably altered.

The Plague, or Black Death, had never really left this country since it had been introduced from the Continent in 1348. It entered England at a port in Dorsetshire and spread over the whole country during the next year.

So conspicuously was 1349 a "Plague year" that in a certain official document of the 23rd year of the reign of Edward III it is alluded to as "in tempore pestilentiae".

The Plague kept on breaking out from time to time until, after its last explosion in London in 1665, the Great Fire cleared it from the Metropolis for ever.

It can easily be shown that the Plague was a feature of the reign of Henry VIII. For we find constant references to that epidemic disease we call "Bubonic Plague", and now know to be due to infection with a specific bacillus, the *Bacillus Pestis*. It came originally *via* Southern Europe from the East, where it still exists. In Florence it was the occasion of Boccaccio's "Decameron".

It was to escape the Plague of 1665 that Milton left London and went to live in a cottage still standing at Chalfont St Giles in Buckinghamshire, where he wrote much of "Paradise Regained". It was, too, on account of Plague that Newton left Cambridge and, returning to his native Woolsthorpe, saw the apple fall that led to such momentous calculations.

Of course efforts were made to combat the pestilence, but with very little success.

The burning even of various kinds of aromatic wood in the streets had no effect. Attention was early directed to the decomposing garbage and to the filthy state of the streets and their open sewers.

Not for about two hundred years after this date (1525) were the scientific principles of ventilation understood, so that the houses, even of the wealthy, where there was no overcrowding, were lamentably deficient in fresh air. There was a most unfortunate impression abroad that smells were in themselves disease-bringing, and that the air was in itself "pestilential" and ought, therefore, at times to be carefully excluded. Thus Erasmus in one of his letters (undated but written between 1512 and 1515) recommends that the windows of the English houses "be made to fit much more tightly than they do, and that there be no access of noxious winds through gaping seams". In 1543 the first serious order to have the streets of London cleaned was promulgated. It commanded the killing of dogs, where to-day we should kill the rats.

It is quite interesting to learn that Erasmus highly disapproved of the English houses, as built with too little regard to the aspect of their doors and windows towards the sun. But until well within our own memory this factor was often not reckoned with by architects.

Of course far more important than all these things was the disposal of the bodies of persons dead of the Plague itself. The burial of the dead had for many years before the time of Henry VIII been a serious problem even in the " no-plague " years. The church-yards attached to the London City churches, never large, were constantly being built over as the need for new houses became urgent. New spaces were from time to time acquired, but these had to be attached to the consecrated houses, the monasteries, and the monks came to regard the fees for the burials as an important part of their income. It is therefore evident that the suppression of the monasteries did not tend to improve the public health, however much it may have improved ecclesiastical morality.

During the whole reign of Henry VIII and as late as 1549 the Black Death was a very real and serious happening. Now it was in Henry's reign that the first version of the English Prayer Book made its appearance. It was confessedly to be in harmony with the new views of the Reformers, that is, it was anti-Papal, but the specific mention of Plague is so interesting to us, seeing that this cannot but be an allusion to the Black Death. The Plague had been a scourge from the year of Henry's accession, when it was in Calais as well as in England. In 1511 the mother of Edward V died of it. In 1513 the Venetian ambassador writes home that in London 200 deaths a day are taking place from the Plague, and that it had reached the fleet. In 1515 twenty-seven nuns died of the Plague within a few days in a convent at the Minories. In 1517 the state of matters was worse, for the Plague was now accompanied by the Sweating Sickness or *Sudor Anglicus,* the third visitation of that pestilence since its introduction by the army of King Henry VII in 1485. In October 1517 the Court was still away from London at Windsor, and by November 16 things had become so bad that the Venetian envoy implored the Senate of Venice to allow him to return, as he thought it was time to " escape from sedition, sweat and plague ". By " sedition ", an Italian writing in 1513 could only have meant the doctrines of the Reformers. On July 22, 1518, the ambassador again asked to be recalled, as two of his servants had died of the Plague, and he himself had had the sweating sickness twice in one week.

In 1529 Campeggio, the papal legate, wrote that the Plague had begun to rage very vigorously. In 1531 the Venetian Ambassador reported 300 to 400 deaths from Plague a week. The summer of 1540 was a sickly one, and a new type of disease, the "hot ague", came to be mentioned; this may have been "the new ague" to which we find a reference in 1537. In 1543 there was a great death (*magna mortalitas*) which lasted so long that the Michaelmas term had to be kept at St Albans.

It is, therefore, perfectly certain that those scholars who composed the first English version of the Litany (1543) had the best of reasons for wishing to be delivered from Plague and pestilence—namely, the presence of the horrors of these all around them.

Now sudden death is one of the characteristics of the Plague. Defoe and all those who have described it have laid stress on this feature. The Plague was sudden in its onset and sudden in its ending. It was the extremely swift course it ran that so terrified all classes alike. In these circumstances it is not at all remarkable that we should find the prayer to be delivered from "sudden death" associated with the prayer to be saved from Plague and pestilence.

In the first Prayer Book of King Edward VI, printed in March, 1549, we can thus perfectly understand why "lightning and tempest" are mentioned in the same prayer as "sudden death", just because they are such pre-eminently sudden occurrences.

In this connexion it is interesting to read in a document, dated July 18, 1551, preserved in the Calendar of State Papers, the phrase, "the extreme plague of sudden death". It is an order from the King (Edward VI) and his Council to the Bishops commanding them to exhort the people to a diligent attendance at Common Prayer so as to avert the displeasure of Almighty God who has visited the realm "with the extreme plague of sudden death"—the very words of the Prayer Book published only two years before. But these measures do not seem to have produced much result, for Queen Elizabeth, on August 1, 1563, had to appeal directly to the Archbishops of Canterbury and York to give orders for a day of general prayer " on account of the Plague " (Calendar of State papers, Elizabeth, Vol. XXIX, 1563, Aug. 1).

The reference to famine in the Litany is equally interesting. The famines of mediæval England were notorious for severity and frequency. Dr Creighton, in his scholarly " History of Epidemics in Britain ", records the occurrence of no fewer than twenty-two great famines in England between A.D. 679 and 1322, or one in about every thirty years. Creighton quotes a well-known proverbial saying of the Middle Ages, " Anglorum fames, Gallorum ignis, Normannorum lepra " (Famine for the English, St Anthony's Fire for the French, and leprosy [or syphilis] for the Normans).

But more than this ; besides famines by themselves and pestilences by themselves, there were " famine-pestilences ", or pestilences accompanying or immediately following upon famines. The pictures of the " rude plenty " of the tables in the castles of England in the Middle Ages may be true pictures—but of what was in the Baron's Hall, not of the inside of the cottages. The memories of the famines were quite vivid enough to cause those who composed the Litany to mention them when they were thinking of the pestilences to which they were so closely related. Thus " famine " comes to be specifically referred to.

Finally, it is interesting to recall that in its long-forgotten origin, the Litany was closely related to pestilence. So far back as A.D. 590, Pope Gregory I, stirred by pestilence that followed on an inundation of the Tiber, ordered the singing of a special Litany, " litania septiformis ". This form of invocation was a hymn-like prayer in which all the following took part—clergy, laity, monks, virgins, matrons, widows, children, and representatives of the poor. The three days before Ascension, Rogation Days, were to be devoted to the chanting of this Litany. Thus, when in church we hearing the priest asking to be delivered from lightning, plague, pestilence, and sudden death, instead of fixing our attention on the inapplicability of it to our modern life, we might remember that that prayer was composed some two hundred years before lightning-conductors had been devised, and while as yet the soil of England was soaked with the poison of Plague, and its churchyards were overflowing with the bodies of Plague victims.

The Black Death was a sudden *death*; it was in terms of no metaphor that in 1549 people craved to be delivered from it. Only sixty years before, the most deadly Sweating Sickness had been a dreadful epidemic in England, so that most of the divines who drew up the Prayer Book of Edward VI could have remembered the first visitation of one of the worst pestilences that ever visited our shores. They had all lived through a reign, that of Henry the Eighth, scarcely a year of which had been free from the Black Death.

It must therefore have been with a peculiar sense of the nearness and the terrible reality of these things that those good men prayed with poignant and personal supplication to be spared a visitation of "Plague, pestilence, famine, and sudden death".

CHAPTER IX

ARE WOMEN MORE SENSITIVE THAN MEN?

AT one time it was the common belief that women as compared with men were more "sensitive".

It was supposed that they had more delicate sensations, a finer "ear for music", a more perfect appreciation of colour, a more discriminating sense of touch, a more delicate sense of smell than had the coarser creature, man.

When these beliefs came to be tested by the rigorous methods of modern physiological psychology, some of them were found to be far from the truth. At the outset we must distinguish acuteness of sensory perception from emotional affectability. The two things have nothing to do with each other. A person may have the acutest hearing and be as cruel as Nero, or he may be colour-blind and yet exceedingly kind-hearted.

Sensations and emotions are quite distinct

aspects of the mind or consciousness, but, popularly, they have been a good deal confused.

Because women were " delicate ", in that they had smaller bones, weaker muscles, and (it was thought) more easily aroused emotions, they were called more "sensitive " ; but if this means having acuter sensory perceptions, the statement requires to be examined critically. If by sensitive, " fastidious " is meant, it would perhaps be a more correct pronouncement.

Let us in the first place consider the senses of smell and taste so closely associated and yet psychologically quite distinct.

We have only five true tastes : sweet, bitter, acid, alkaline (soapy) and salt ; and these never leave us even when we have a bad cold-in-the-head.

For a cold-in-the-head blocks up our nose, where are found the end-organs for smelling, and so what really vanish are our perceptions of odours and flavours, *smells* in fact, and not tastes at all.

Now women perceive the five true tastes just as perfectly as men do ; but the odours they do not.

The enjoyment of food and cooking is

very largely a matter of the appreciation of these odours and flavours; so that a gourmet is a person who enjoys his food, fruit, wines, etc., because his sense of smell is acute, not because his five true tastes are in any way different from those of other people.

It was long ago pointed out that there is no feminine form of "gourmet", the French apparently assuming that the person who especially appreciates good cooking is a male. Of the word "gourmand" there is a feminine form, a gourmand is merely a person who eats largely.

It is owing, therefore, to a deficient sense of smell that women are not so critical about flavours, odours, "bouquets" of wines, etc., as men are. Thus, too, it comes about that it is always men who are tea-tasters or professional connoisseurs of wines.

It used to be true that the flavour of the tea was better in men's clubs than in women's.

This general impression of the inferior delicacy of the female sense of smell was fully confirmed in an elaborate research done some years ago by two American professors who examined the olfactory sense in a large number of men and women. In this connection we may recall the fact that it is women who are apt to overdo the use of scent, nor does this seem to be a recent failing, for Saint Clement of Alexandria wrote: "Let a few unguents be selected by women such as will not be overpowering to a husband."

The next sense in which women are deficient is the muscular sense or sense of movement and of resistance to muscular effort; it is sometimes called "the sixth sense".

The muscular sense tells us, for instance, just how hard to strike a nail, how much effort is required to twist round various sizes of screws, how heavily we may press on a spring, how powerfully turn a key, and so on.

In particular, as a teacher of practical physiology, I used to note that in those exercises where a long, delicate, easily broken lever of straw with a paper writer on it had to be pressed against a recording surface with just the right amount of pressure, it was always the men who first acquired dexterity in this manipulation. The women broke more of these straw levers during the session, for they pressed the lever either too tightly against the smoked surface or not tightly enough.

For the third and last sense in which women are deficient, let us take the sense of the perception of cold.

It is men who first perceive or complain of a draught. It is a matter of common observation how many fewer clothes women can wear with apparent comfort than can men. At one time this was attributed to the larger amount of fat under the skin, which most women have as compared with men; and while in some cases this may be part of the explanation, it will not account for the thin, spare girls of to-day in their flimsy dresses being quite as comfortable as their plumper sisters of a generation ago.

Cold does not seem to produce such impressions of discomfort in women as it does in men.

And this agrees with that other well-attested fact that women endure pain better, in some cases much better, than men, for the perception of cold and of pain are closely allied.

Thus summing up we may say: women have a less acute sense of smell, of cold, and of muscular sensation.

CHAPTER X

BLAMING THE AIR

"WHEN in doubt" blame the air seems to have been the precept followed by such of mankind as tried to get at the origins of certain things.

Not once or twice in the course of the history of discovery the air has been blamed for ill effects when the source of the trouble was something quite different. There are at least four outstanding examples of this curious mistake: the first when the idea got abroad that malaria was caused by the air or vapours from marshes; the second when night air was thought to be injurious; the third when the oxygen of the air was believed to be the source of the origin of living "germs"; and the last when the air was blamed for the "going bad" of wounds.

The belief that malaria was, as the name indicates, due to "bad air" (Italian, *mal aria*) is a very old one. But we can easily see how it arose, for the source of the scourge seemed to be in the neighbourhood of marshes and damp places. The invariable association of damp air and malaria seemed too obvious to be weakened by the question: How can air, damp or dry, be in itself the source of any disease, seeing that air is the very breath of life?

So firmly was the malign influence of marshes believed in that the term "paludism" (from the Latin *palus*, a marsh) was given to it. We now know that all that marshes have to do with malaria (ague) is to be the places where the mosquitoes breed that carry the parasites of that disease.

Before the discovery of the microscopic causes of a vast number of diseases—most of them made within living memory—it was so very easy to mistake one thing for another as the source of some particular kind of mischief. The search for causes, which is science, is notoriously difficult, and before one has got a clue, it is very easy to be misled. Damp air seemed always associated with ague, therefore as vapours and nothing else could be *seen*, the damp air was credited with the mischief. It was entirely false; damp air may predispose to illness, but neither vapour nor air of itself is the specific origin of any disease.

The next example is the notion that night air is bad. Until quite recently this was firmly believed. As in most errors, there is a germ of truth in it. During the night, living vegetation exhales carbonic acid gas, and this same substance is one of the waste-products of animal breathing. Carbonic acid gas has long been regarded as poisonous to animals. It is the old story of calling a thing a poison, and then regarding it as such in all dilutions.

Alcohol is a "poison"; above a certain concentration it is thoroughly injurious; below a certain dilution it is anything from a stimulant to a negligible quantity. The idea that night air is bad arose from an injudicious and, indeed, erroneous remark of the Dutchman, Ingenhouz, who in 1779 discovered the fact that during the hours of darkness plants exhale carbonic acid gas. So they do, but not in such a concentration so as to harm either men or animals.

The third example of supposing that the air was the source of some particular thing when that was something totally different is in connection with the origin of life itself.

It is not so very long ago that the great controversy whether life can arise from the non-living was closed. The dispute in its final stages became narrowed down to the problem whence arose the living things that undoubtedly were found growing in carefully sealed-up flasks of organic material put aside for a day or two in a warm place. Before the microscope was applied to the problem, it could not be known that "the life" in these flasks consisted in myriads of extremely minute organisms—the micro-organisms of putrefaction—having gained access to the putrescible material and continued to live in it.

It all looked as though the air was to blame, for if the flask of broth was left open to the air, life appeared in the broth; and even if the air in the flask was corked up there were still evidences of vitality in the organic contents. The air seemed to have been the cause of the living things.

But in reality it was not so; after a long and bitter controversy it was shown that air from which everything had been filtered off through cotton-wool, or which had been passed through a red-hot tube, would not give rise to life in the flask.

It was the mistaking of one thing for another—mistaking the carrier of the living things for the things themselves. The oxygen of the air does indeed support life, but it can no more give rise to life than can any other gas or mixture of gases.

The fourth attempt to blame the air was in connection with the introduction of antiseptic surgery. It had for long been a familiar observation that wounds exposed to the air—and more particularly to indoor air—" went bad ". This exposure to the air—especially hospital air—as we now know, meant exposure to dust and the invisible microbes; but before Lister applied the microscope to this problem, the true cause of wounds " going bad " was not suspected. The air as usual was blamed for the frequently disastrous condition of open sores.

With infinite patience Lister showed that it was not the air but what it held that was to be blamed, not what it was but what it carried that was objectionable. Under the microscope Lister beheld the minute cellular forms which were causing the pus to putrefy, and so once again the air was exonerated.

We now know that pure air is very beneficial to wounds; like all living things, wounds are the better of fresh air. On high mountains and in the germ-free air of the ocean surgeons know well that wounds heal rapidly.

Lastly, fresh outside air was at one time actually supposed to be detrimental to people suffering from consumption of the lungs; so these unfortunate ones were shut up in hot rooms from which fresh air was excluded. To leave phthisical people sitting out-of-doors, as is done to-day, was not so very many years ago regarded as a species of murder.

As far back as 1840, Dr George Bodington, of Sutton Coldfield, published his "Essay on the Treatment of Pulmonary Consumption", in which he pointed out the advantages of dry, cool air for closing and healing cavities and ulcers of the lung. But like all new things it was resisted: Dr Garrison, in his "History of Medicine", tells us that "Bodington's theory was so roughly handled by the medical critics of his day that he was discouraged from carrying it into practice to any extent".

The pioneer painfully smooths a path which others who follow him tread in comparative comfort.

CHAPTER XI

HEATING AND VENTILATING THE HOUSE

Two Age-long Problems

It would be safe to say that most people confuse ventilation with heating. They assume that because a hall, room, or railway carriage is cold it must be well ventilated, and, conversely, that because any of these is warm it must be badly ventilated.

But a room may be warm and well ventilated, or it may be cold and badly ventilated. From the standpoint of physics the two problems—how to heat and how to ventilate—are quite distinct.

To heat a room is to bring the temperature of the air in it to such a degree that it is not "chilly" (unpleasantly cool) to a person in indoor clothing at rest in that room.

To ventilate a room is to provide for the ingress of pure out-of-door air, and the egress of used air *without* a *draught*.

A draught is air in motion taking heat from the skin so rapidly that it is unpleasant, and may be dangerous to health. Any one can ventilate a room by creating a draught; all he has to do is to open the door and the window, when the air is more or less rapidly blown out of the room (perflation). But, except for a short time, no one could endure that draught; only the most fanatical fresh air "fiend" would enjoy it.

We are not contemplating the pleasant draughts so desired in the tropics, and occasionally during an English summer, but the ordinary, every-day rush of low temperature air chilling the skin and often predisposing to definite illness. It would seem that women are much less sensitive than men to draughts and cold generally. Their cold-perceiving nerves seem less acute. Thinly clad, they can endure without apparent discomfort degrees of cold that to the mere man would be quite intolerable. Many of them, as we know, enjoy sitting in a railway carriage facing the engine with the full blast of the cold air rushing on to them.

The pleasantest room is that where the incoming air is fresh, directly derived from out of doors and somehow heated to a temperature of about 65° F. This will not disagreeably cool the skin of a healthy person inactive in that room.

Now in the vast majority of cases, the ordinary open coal fire does not fulfil these conditions, for although it does change the air efficiently, it heats it very poorly. It is estimated that from 60 to 90 per cent. of the heat of the domestic fire rushes up the chimney, the remainder being radiated into the room. This radiated portion warms any solid objects it may encounter, and after a long time the air is warmed by contact with these. But all the time the air so slowly warmed is tending to flow up the chimney.

Thus it is that a person facing a fire may have his face and hands scorched, while his back is kept quite cold by the draught. In other words, the old-fashioned English open fire ventilates well from the first, but heats poorly later on.

Its other drawbacks—the cost of coal, dust, cinders, smoking chimneys, the cleaning of grates and flues, and the occurrence of fogs in cities—are, in the well-known phrase, " too familiar to mention".

Now it is supposed that the Americans and Canadians have solved the problem of heating a building by the method of having hot water or steam circulating continuously through metal pipes called " radiators " which yield their heat directly to the air of the room. It is quite true that the air can be heated in this way to temperatures far above 65° F., to 70°, 75°, or 80° F., as may be desired.

But most people from Great Britain find these higher temperatures distinctly disagreeable, largely because the air so heated is not also being removed. Naturally the coils provide neither for the incoming of fresh air nor the outgoing of used air.

A great many bedrooms on the American continent are not provided with fireplaces, flues, or other apertures to afford escape of the tainted air, which, so long stagnant, acquires a " stuffy ", disagreeable odour.

On many winter days the windows, which are often " double ", cannot be opened, because the very cold air rushing in would freeze the water in the coils.

It is true that in some of the better-constructed houses the two systems—open fires and hot water pipes—are employed, and little fault can be found with such an arrangement.

But in a very large number of Canadian

houses the bedrooms are not provided with open grates, and hence the anomaly of the prevalence of tuberculosis in a country of vast unoccupied spaces, for tuberculosis is pre-eminently a bedroom disease.

The system of central heating as usually installed does not provide for the changing of the air. Architects know well that if we provide for the exit of the used air, the incoming air will look after itself; if, now, we warm this latter, then we have solved the age-long problem of heating and ventilating.

However well warmed the ordinary brick house may be, it is rather wasteful of heat, for brick allows heat to escape faster than stone does. Wet bricks are particularly bad in this respect.

One of the latest devices for keeping houses warm is to use hollow bricks; the air inside them, being a bad conductor of heat, conserves the heat of the house to a very considerable extent. Air-filled spaces are also used in the latest type of house, where the walls are constructed of hollow concrete blocks.

The utilization of this principle does not seem to be any discovery of our own time, for the Romans certainly used hollow bricks in their buildings.

We may see some of these hollow bricks amongst the ruins of the Roman baths at Bath, and amongst the *débris* of camps and villas erected during the Roman occupation of Britain.

The Romans seem to have understood the economical use of heat, and also to have known how to heat their houses by a method much superior to that of the open fire. They heated *the floor first.* They built the floor over a space, the hypocaust, so that hot air from a fire or furnace could be made to traverse this space below the floor. Thus the room was heated from the floor upwards; and it is well known that if our feet are warm we do not feel cold, but if our feet are cold, then no matter how warm the rest of us may be, we cannot feel comfortable. Fine examples of hypocaust have recently been unearthed at Folkestone.

This warming of the floor from below is obviously quite different from the sending up of hot air through gratings in the floor, as we often see done in cathedrals and other large buildings. This is called heating by convection. It is a wasteful method, for the heated air rushes up past the people towards the roof, so that the atmosphere far above their heads gets heated first.

In the new cathedral at Liverpool the Roman method of heating the *floor* by hot air has been adopted.

A system of heating has been devised which has overcome most of the objections to all other systems. It consists in embedding the hot water pipes in the floors, walls or ceilings, so that these large surfaces, soon becoming warm, radiate their heat into the rooms.

In this—the so-called "panel" system—the air itself is not used as a medium of heat transmission, so that it is possible to ventilate the building without an undue loss of heat.

These coils can be placed below floors of every kind of material—wood, marble, slate, tiles or mosaic work.

The stuffy smell of warm rooms is due to the volatilizing of organic dust; in the panel system this dust does not become disintegrated to anything like the extent that it undergoes over exposed hot water coils. Thus it is that buildings heated on the panel system have so little of the stuffy smell characteristic of most hot rooms.

Heating by gas is, of course, much cleaner than by coal, but the gas-fire comes under the same condemnation as the coal-fire, in that it is better as a ventilator than as a heater of a room, and this only when it is connected with a properly constructed flue. The idea that gas-fires are unhealthy is quite a mistaken one: provided that the products of combustion are removed, the gas-fire is as " healthy " as the coal fire.

One reason why the British find the air of rooms heated by hot pipes disagreeable is that such air tends to become much too dry. In too few instances do we find the " harmless necessary " vessel of water provided. The water evaporating into the air makes it comfortable for breathing. Air that is too dry irritates the throat and lungs, inducing the catarrh from which, as is well known, so many of our American friends suffer.

The latest source of heat is electricity; in that it does not consume the oxygen of the air, it is an excellent one, and it may be quite inexpensive if it be produced by water power instead of the combustion of coal. Electric heaters can be employed to heat the incoming air in a much more convenient fashion than can coal or gas. Better still, electrically-heated wires, etc., can be buried under the floors, in the walls, or in the ceilings, thus allowing these large surfaces to become true radiators of heat.

To conclude, then, a perfect system of heating and ventilating in winter is one which provides for the inflow of pure, out-of-doors air warmed to at least 65° F., and subsequently, if necessary, moistened, but not saturated with vapour, the used air being finally discharged into the ocean of the atmosphere.

CHAPTER XII

PHYSIOLOGY AND "VITAL FORCE"

EFFORTS in the past to account for the mysterious powers and properties of living beings have in the main consisted of bringing in some supra-sensible, immaterial, ultra-cognizable "principle" or "entity". This has been known under different names as time went on, but they have all referred to the same thing—psyche, pneuma, the anima sensitiva of Van Helmont, the life principle of Stahl, the Bildungstrieb of Blumenbach, the vita propria of Bordeu, the vitalis agens of Barthez, the Lebenskraft of Reil, the entelechy of Driesch, the elan vital of Bergson, and the biotic energy of Moore. It is of the essence of "vitalism" to explain life in terms of the less known, to account for the properties of a living being by the indwelling activity of an entity, agent, or force which, by its very metaphysical nature, is inaccessible to human investigation.

To assert that each organ, tissue, and cell is as truly alive as is the whole organism is not " vitalism ", but a correct physiological observation, for the isolated organ—heart or liver—can perform its vital functions long after separation from the body to which it belonged; and as Ross Harrison and Carrel have proved, the cell taken from its tissue can live for years in its morphological isolation. To say these organs have an independent life (whether it is said in English or in Latin, *vita propria*), is to report an observation; to say that life is due to a " life force " or " entelechy " is to state a theory.

If the vitalists had always been careful to admit that their view was only a theory of life, much acrimonious discussion would have been avoided; but when they declared that theirs was the only right view, their virtual claim to omniscience failed to arouse in many minds the enthusiasm expected. These unconvinced people tried to explain life in terms of what they knew at least a little about—the non-living world around them; and they were in consequence called "materialists" and their theory "mechanistic". Now theirs is a theory no less than is the vitalistic. The materialistic theory is that the observed known—the properties and behaviour of living organisms—may be explained by applying to them our knowledge of the laws and properties appertaining to the non-living world. It is an explanation of the known in terms of the more known rather than the less.

The term "materialist" ought not to be used as one of reproach; the materialist is almost always a sincere searcher after truth, who, starting from his colleague the physicist's knowledge of the properties and behaviour of *non*-living matter, attempts to apply these to the behaviour of living matter. He finds that many of the laws that hold good in the world of the non-living seem to be equally applicable to that of life. In particular, he finds that vital heat, for example, is not in its essence different from heat of non-vital origin, that the great generalization of the conservation of energy holds good for the mammalian body, that "vital" processes are accelerated by a rise, and retarded by a fall, of temperature exactly like "purely" chemical reactions in a test-tube. He is able to say in the language of his chemical colleagues that living matter (protoplasm) behaves in many respects like an irreversible colloidal emulsoid hydrosol.

The materialist found, as a matter of fact, that so many vital activities seemed to be the outcome of the operation of laws already proved true for the non-living universe, that he finally made so bold as to assert—and here he made the first mistake—that the mystery had vanished, and that protoplasm, chemically speaking, was only an excessively complicated form of matter. To this view Loeb committed himself. He wished us to believe that he had proved that the dividing line between the non-living and the living had been removed, and that we might pass by a number of gradations from physico-chemical simplicity at one end of the series to great physico-chemical complexity at the other. Somewhere on the way one passed from the non-living to the living. Here the materialist went beyond what his premises allowed him; in this he was rash; but rashness inheres in the enthusiasm of youth, and biology is a very young science.

But the vitalist, too, had always been rash, for he had the boldness to go beyond experience. He had asserted, for instance, that urea and sugar could never be made without the agency of life, because he had never found these substances anywhere except in living animal or vegetable tissues. When, therefore, in 1828 Wöhler made urea in a test-tube, and in the 'eighties of last century Fischer synthesised sugars, and when indigo was produced that had never seen a plant, and when the hormone adrenalin was constructed artificially, vitalism received a set-back, and materialism a corresponding encouragement. The anti-vitalistic view was further strengthened by the discovery that many ferments, solely the products of life, dealt with their "substrates" exactly like the inorganic catalysts so well known to chemists. Finally, when Loeb, by altering either the density or composition of sea-water, caused the unfertilized eggs of the sea-urchin to begin to undergo development, the days of vitalism seemed numbered.

The materialist has, indeed, shown us how the plant is able to synthesise a carbohydrate from carbon dioxide and water through the stage of formaldehyde; but he himself is very far indeed from making a scrap of nucleated protoplasm, and until he does that, since we know no life apart from nuclei, it seems exceedingly improbable that he will be able to synthesise any kind of living substance. The materialist goes far beyond his observations when he maintains that the laws of the non-living world are the only laws which operate in the sphere of the living. It is one thing for them to operate there, it is another for them to operate *alone*.

Life and living things are *sui generis*. It is better to admit at once that, so far as our experience goes up to this hour, life is unique. Prof. J. A. Thomson gives the following features of the uniqueness of living matter thus (Gifford Lectures, St Andrew's, 1915-6)—

1. Its capacity for enregistering experience which, in the hereditary relation, is an expression of physiological inertia.

2. The self-maintaining tendency of the organism.

3. Its variability, or the capacity to give origin to the new.

Living matter can do what no non-living matter can—assimilate material wholly unlike itself; evolve from a minute and relatively homogeneous speck into an obvious and heterogeneous organism; reproduce its kind either by casting off buds or by gametes; pass through a life-cycle of irreversible stages, from infancy to youth, maturity, senescence, and death.

So far as we can judge, the higher animals possess a variety of energy, nerve-energy, which also is *sui generis*. Finally, in the realm of the living we encounter those phenomena called "subjective", the world of consciousness with all its modes—sensation, emotion, volition—a world which, so far as we know, never exists apart from living matter, and is absolutely unique.

Certain critical vitalists sometimes blame physiologists for investigating living things by the methods of chemistry and physics. The late Sir James Mackenzie wrote (*British Medical Journal*, March 1, 1924): "Physiologists place the organ in artificial conditions, employ artificial stimulation and obtain artificial results . . . such as the all-or-nothing law".

Now this criticism is rather hard to bear, because we have no methods or apparatus other than these "artificial" ones. Although the biologist is investigating phenomena that are *sui generis*, he has unfortunately no apparatus which is adapted exclusively to biological use. The physiologist uses test-tubes, retorts, induction-coils, and galvanometers because there are no two kinds of instruments, one for biologists, and one for physicists. It is rather hard that, having been forced to use these things and having by so doing obtained results valuable to medicine, he should be stigmatized as a materialist.

When we speak of " the mechanism " of gastric digestion, we merely use the expression as a conveniently short one for all those processes which are involved in the vital manufacture of pepsin and hydrochloric acid from neutral blood, their separation through and by the mucous membrane, and their solvent action upon food in the interior of the viscus. Nothing more " mechanical " is meant than this, and this is not mechanical at all. Only the last process, the solution of the protein, can go on *in vitro* ; only life can manufacture pepsin from blood, or an acid from alkaline or neutral salts, and separate these without digesting the wall of the stomach in the very act itself. It is most unfortunate that anything to do with a machine should be mentioned in this connection.

Because the stomach " works " rhythmically and predictably we may call it a machine for turning out pepsin from blood, and liken it to a machine for turning out (say) newspapers, but the secretion of pepsin is not mechanical, nor is the output of newspapers vital. Processes that closely resemble each other are not necessarily identical. In a decerebrated animal the same predictable reflex action can be obtained time after time from the same stimulus, and we may speak of the inevitableness of "the mechanism" of reflex arcs, but in no other sense is the spinal cord a machine.

Non-vitalistic physiologists are blamed not merely for speaking of organs and organisms as machines, but also for daring to study organs and systems by themselves, whereby, it is alleged, they have lost sight of the wonderfully co-ordinated activities of the animal as a whole.

It would seem that the physiologists can never do anything right; at a time when nothing was known of the heart, or of the liver, or of the retina, what else could the investigator do than confine his attention to one thing at a time? Some of us, indeed, are trying to take a comprehensive synthetic outlook over the whole field of vitality, but that could not have been done by the pioneers, who could only push their way slowly into an unknown territory.

One would infer from the obituary notice of the late Sir James Mackenzie in *The Times* (January 27, 1925) that physiology is bankrupt. Sir James is quoted as having said: "When this" [his own view of physiology] "is realized, the whole of the physiological interpretation of the functional activities of organs will have to be scrapped. This is one of the results that I anticipate".

The physiological methods by which were discovered the localization of cerebral function, reciprocal innervation, the endocrine function of the adrenals, the thyroid gland, and the pituitary, by which insulin was isolated and adrenalin synthesised, cannot be discredited.

The experimental work of Schafer, Bayliss, Starling, Sherrington, Pavlov, MacLeod, and the Hills is not only not to be scrapped, but is frankly to be recognized as the logical modern basis of practical medicine. The researches that led up to electrocardiography, with which Sir James Mackenzie himself was so conspicuously associated, are another illustration. From some of his expressions the incautious reader might suppose that current physiology was useless as an introduction to medicine, and that the secret of success in diagnosis and treatment consisted in the reinstating of a "vital force". Mackenzie, writing in the *British Medical Journal* (March 1, 1924), thus expressed himself: "The reason for the lack of progress in respect of [knowledge of] living matter has been the absence of a knowledge of a vital force!" The whole tenor of this paper is that we must distrust the results of "artificial" stimulation of tissues because the effects of these are not normal, are not the same as those produced by the "vital force". The experimenter, in fact, comes under a heavy condemnation. We are further told that because we do not better understand pain and why certain reflexes occur, this "vital force" must be postulated.

The term "impulse" plays a large part in this neophysiology. We are told: "Where a cell discharges its energy in the shape of its peculiar [? particular] function, it at the same time discharges an impulse". We must know, therefore, where in the conceptual scheme of things, which has served physiology so well for the last thirty years, we can place this all-important "impulse".

As regards protoplasm three concepts are fundamental—(*a*) its affectability (irritability), (*b*) the stimulus that may operate upon it, and (*c*) its own response. The behaviour of a living efferent nerve may best illustrate the relations between the members of this biologic trinity. The neuroplasm, in virtue of its possessing the property of responding to a stimulus (affectability), and having received a stimulus, responds by giving rise to an excited state which travels down the nerve as the impulse to the effector organ at the periphery, say a muscle, which twitches when the impulse impinges on it.

Let us suppose that the stimulus is a blow on the nerve, or a pinching of it, seeing that the electric stimulus of the induction coil was regarded so unfavourably by Mackenzie. Of course the existence of this impulse is an inference from what happens: a nerve receives a blow at one end and a muscle twitches at the other; something must have passed down the nerve, and that something we have for a long time called the impulse. It is said to travel down the nerve by reason of the nerve possessing conductivity.

The other indication that something is passing along the nerve is that if a galvanometer be substituted for the muscle, then when a blow is given to the nerve, the galvanometer records the presence, for a very short time, of an electric disturbance in the nerve. We infer that this electric current is an outward and visible sign of the existence of the invisible nerve-impulse. But just here a very important conclusion is reached, namely, that the impulse in the nerve on arriving at the muscle is for the muscle a stimulus to it to "contract"; the impulse in the living nerve can, then, an instant later, be the stimulus for the living muscle, the response of which is a state of shortening or contraction.

We infer that in the intact animal the nerves are conveying impulses normal or homologous, exactly similar to our laboratory ones, because if a nerve *in situ*, for example the phrenic, is connected to a galvanometer, electric currents in this case also are seen to pass through that instrument. These natural (normal) impulses must be the natural (normal) stimuli for the muscles *in situ*.

All this is very elementary; but it is evidently necessary to restate it because it accounts for everything that Mackenzie observed without calling in the aid of a "vital force " at all. We fail completely to see where and why this force needs to be invoked, and why physiology is bankrupt if it is not so introduced.

If this " vital force " is another name for the nerve-impulse—and it can scarcely be synonymous with either " stimulus " or " response "—then it is a superfluous term. If it is not a synonym for any of the three, then it is some fourth thing for which apparently there is no place in the scheme which has served biology so well in the past. The neo-physiologist may reply: it *is* a synonym for the nerve-impulse, but it comprehends " impulses " in all other tissues. Mackenzie's own words, in a paragraph headed " The *impulse* a *vital force* ", were: " An impulse is the product of cell-activity, it can only be conveyed by living structures and acts by stimulating cells to discharge their function . . . it differs from all other forces ".

If the impulse in this passage be confined to *nerve*-impulse, no fault can be found with it, for the nerve-impulse is the product of (nerve) cell activity, it can be conveyed only by the living structures of the nerve, it acts by stimulating (muscle) cells to discharge their function (of shortening), and it differs from all other forces in that, as a nerve-impulse, it is *sui generis*. But why declare that unless we call this impulse a "vital force", physiology is to be discredited ("scrapped")?

We do not need an additional name for the nerve-impulse; if we are forced to give it a new name, why give it one so redolent of obscurantist animism?

It is, however, quite clear that by "impulse" Mackenzie meant something that was active in all the tissues, for he speaks of *cells* in the widest sense. Now what are these impulses in tissues other than neural? What impulses are there in muscle, connective tissue, gland, fat, bone? In muscle we have states of contraction; and we can *call* them "impulses" if that will rescue physiology from the scrap-heap, but one fails to see what is gained thereby. As for "impulses" in connective tissue, fat, or bone, we have no evidence of them. In glands, doubtless, some states of excitation can travel (slowly) along, but again we see no benefit from calling them "impulses".

It is difficult to see, therefore, why physiology is to be declared as proceeding on a totally wrong road. It may frankly be admitted that there is in physiology more than is dreamed of by the mechanistic philosophy. The laws of matter that has never lived have failed so far to account for certain facts, for example, about absorption both from intestine and bladder, and for certain facts about urinary secretion. It has been shown that living membranes act very differently from dead ones.

The materialistic view of life has failed signally to account for certain realities of our experience of which consciousness is the group name. Huxley himself made his bow to consciousness and waived it away as an epi-phenomenon. For the materialist, consciousness cannot be a cause of neural activity; nor can states of body affect mind, for mind is an illusion. The mechanistic theories are incapable of throwing any light on the central fact of experience, the permanence of personality amid the flux of matter.

Prof. Haldane has well said that "living" and "mind" are not reducible to simpler terms; they are the axioms of biology; and this thinker firmly believes that physiology is being retarded by mechanistic conceptions which deprive us of a complete view of life. As Prof. J. A. Thomson has said:[1] "We need new concepts such as that of the organism as a historic being which has traded with time". "We need these new concepts because there are new facts to describe, which we cannot analyse away into so-called simpler processes."

The most reasonable position to assume as regards vitalism is to insist that there is no compulsion for the biologist to be either a "materialist" or a "vitalist". It is quite open to him to say that as he is dealing with an order of things that is unique, with existences that are *sui generis*, and that as his science is so young, he is not yet in a position to dogmatize and declare that *qua* life there are no categories beyond those which the physicists and the chemists recognize.

Of all the many wise things Prof. Thomson said in his Gifford Lectures at St Andrews (*loc. cit.* vol. 2, p. 147), this is surely one of the wisest: "We regard the question as one of the many *false dichotomies* with which man in his search for clearness has been led astray".

[1] "System of Animate Nature" (Williams and Norgate, vol. 2, p. 160).

CHAPTER XIII

THE PLACE AND POWER OF THE NERVOUS SYSTEM AND ITS RELATION TO THE MIND

THE nervous system may be studied under all of the following aspects, namely, in its development both in the individual and throughout the animal kingdom, in its naked eye appearances, in its microscopic structure and finally in regard to its functional activity in health and disease.

The unaided eye can see that it consists of a long central axis called the spinal cord within the vertebral column becoming much thicker as it enters the skull where it is known as the Medulla Oblongata. This part is continued along the base of the skull and is, in man, completely covered over by a great development of what is called the grey matter of the cerebral hemispheres. In the central nervous axis arise all the outgoing or motor nerves of the body, 9 pairs from origins within the skull, and 31 pairs from the spinal cord. Similarly all ingoing or sensory nerve-fibres traverse the central axis on their way into the nervous system.

Under the microscope the "grey matter" is seen to consist of multitudes of nerve-cells, minute living units possessing all possible shapes from a simple sphere to a most complicated figure. The most elaborate forms are in the grey matter of the surface of the hemispheres, a portion called the cortex cerebri. This is significant, for we shall see that here, where the structure is most complicated, the functions are of the highest order. The fish, for instance, has no cortex cerebri: its "mind" is of the simplest.

Every nerve-fibre (white matter) takes its origin in a nerve-cell: there is no such thing as a nerve-cell without a related fibre or a fibre without a "trophic" nerve-cell. Indeed, fibres die if cut away from their related cells. The whole structure, cell-processes and fibre, is now known as a "neuron".

The nerves may be functionally divided into two great groups—the efferent or those arising in the central nervous axis and ending in the tissues, and those which do just the reverse. The efferent or motor nerves supply the body-muscles including the heart, the muscles of the blood-vessels, the iris of the eye and the muscle of the intestinal tract; they supply the glands of the mouth, the stomach, the intestines, and the great gland, the pancreas, and finally they supply the kidneys and the reproductive organs.

The second great group of nerves is the sensory, those arising in the skin and in the end-organs of the special senses including those of muscles. A great many nerves, which come from the internal organs, are not related to any definite sensation but are only occasionally in activity as nerves of pain. Nerve-fibres are exclusively conductors; they originate nothing, they only conduct nerve-impulses either outwards to tissues or inwards to the central nervous system. The essential structures in the central nervous system are Nerve-Centres—anatomically larger or smaller masses of nerve-cells.

A "centre" is a group of nerve-cells specialized to control or supervise some definite physiological activity in the body. This presiding over a function is called "innervation", that is, being attended to by the nervous system. As a definite example we might take the respiratory centre in the Medulla. At a certain definite rhythm this centre emits salvos of nerve-impulses to the diaphragm and the other muscles of breathing. In man in health this rhythm is from 18 to 20 in the minute.

The centre, which is bilateral, occupies a relatively small region in the Medulla, so that a quite circumscribed injury here will at once arrest the breathing. It is the sudden destruction of this centre which is the cause of death by hanging. Although the respiratory centre seems to be working all by itself or "automatically", as we call it, a little study will convince us that the centre is accessible to influences from widely separated regions of the body.

In the first place, if we cut the great nerves coming up from the lungs to the respiratory centre, we shall find the rhythm of breathing to be greatly altered. The animal now breathes very much more slowly, although correspondingly deeply, from which we infer that, in the intact condition, the movements of the lungs themselves are in some way able to accelerate the rather slow inherent rhythm of the centre for breathing. But further, impulses from the skin and from the internal organs can also influence it. A dash of cold water on the skin will cause a gasp followed for a short time by a stoppage of the breathing, while severe abdominal pain will alter the breathing to quite a shallow type.

Let us next consider a far more familiar experience. If we so wish it, we can voluntarily hurry up the rate of our breathing, or on the other hand we can slow it or stop it (for a time) altogether. In physiological language this is that from the cerebrum, volitional impulses are stimulating and inhibiting, respectively, the functional activity of the respiratory centre. Besides volitional influences, the emotions are equally potent in affecting the centre for breathing. Pleasant emotions, particularly in children, can accelerate the breathing; while powerful and terrifying conditions can arrest it altogether, as the phrase goes, " it quite took my breath away ".

Finally the chemical composition and the temperature of the blood traversing the centre are both able to influence the centre directly. Increases of the carbon dioxide and of the heat of the blood increase the depth and the rate of the breathing.

Thus we may infer, surveying all this evidence, that the nervous system is for the purpose of bringing rapidly one portion of the body into relation with another, in fact co-ordinating them. Some other centres in the Medulla and adjacent part of the central axis are for the control of the heart, of the diameter of the blood-vessels, for the production of saliva, gastric juice, and pancreatic juice, for the movements of the stomach and intestine, for vomiting, for the movements of the iris (pupil), and for perspiration. The centres for muscles of articulation in the lips, cheeks, tongue, palate and larynx are also found in this region of the grey matter of the central nervous axis.

Broadly speaking there are two great types of activity in the nervous system, that of the reflex action which does not require consciousness for its performance, and the type where consciousness, either as sensation, volition or emotion, is of the essence of the action.

The reflex is the primitive or fundamental type of action, the earliest both in the life history of the individual and of the animal series.

A reflex action on the structural side involves a sensory nerve, a centre and a motor nerve, such a path being called a " reflex nerve-arc ".

The newly born child is a bundle of reflexes and automatic actions. Put to the breast, it sucks reflexly, and will do so even though its cerebral hemispheres are absent ("acephalic monster"). Reflex actions are elicitable from every portion of the central nervous axis.

The contraction of the pupil when light enters the eye is a reflex carried out from an anterior portion, flushing from heat or blanching from cold is carried out at a lower level, while the jerking of the feet when the soles of the feet are tickled is a reflex concerned with the lowest part of the spinal cord. Our reflexes are literally too numerous to mention.

One characteristic of the reflex action is that the will has no control over it.

When grit gets into the eye, the " eye " waters, that is, the lachrymal gland secretes tears in spite of all our desires to the contrary; when the dentist probes the gums with one of his instruments, the saliva flows in spite of our urgent wish to stop it.

The reflex action is outside our control; we are not responsible for it, it occurs within us but we are not accountable for it ; it is physiological activity but not "behaviour".

The other group of actions besides the reflexes is that of those activities of the nervous system which *are* related to consciousness.

The cerebral hemispheres are undoubtedly the physical basis of consciousness ; when they are active, consciousness is present, when they are inactive, as in sleep and chemically-induced anæsthesia, consciousness vanishes. The proofs that the cellular elements of the cortex cerebri are the physical substratum of the mind, the material basis of psychical states, are derived from experimental physiology, from a study of the development of the nervous system and from clinical medicine and surgery.

From the remotest antiquity, the powers of the mind have been believed to be related to the brain within the skull. But it is only quite recently in the history of scientific discovery that specific proof has been forthcoming. Complete removal of the brain (which is possible in some animals) reduces them to a state of utter mindlessness ; they become inert automata, they feel nothing, they desire nothing, they originate nothing.

It is actually only a little more than fifty

years ago that experimental proof was given that the cortex cerebri *is* the physical basis of sensations and of the origination of all bodily movements. The man who first proved these things was Sir David Ferrier, F.R.S., a Scotsman, long Professor of Neurology at King's College, London. Having (under anæsthetics) exposed the living cortex in one of the higher apes, Ferrier applied the interrupted electric current to the surface of one hemisphere and at once obtained movements of muscles on the opposite side of the body. These were of such a kind as to suggest voluntary actions, had the animal been fully conscious. This representation in the brain of the muscular movements of the body was found to be surprisingly detailed. The converse experiment showed that excision of the area "for" a group of muscles brought on paralysis of that group. Naturally these "motor areas" are often called "Ferrier's areas".

Ferrier also investigated the central representation of the senses; and in spite of the manifest difficulty of interpreting the signs of an animal's sensations, he succeeded in discovering the localization of the centres for sight and for hearing.

There is no doubt at all that the sense of vision is located in the occipital lobes at the back of the brain, and the sense of hearing at the sides below the temporal bone. Electrical stimulation of the occipital lobes makes an animal turn its eyes about and evince all the signs of seeing something; while stimulation of the temporal lobes makes it prick up its ears and turn its head as though it heard something and wished to hear it better. Human experiences have fully corroborated animal experimentation.

Thus if a tumour presses on the visual centres, the patient may in consequence have visual hallucinations. When as the result of an injury or a wound in war, the occipital lobes are blown away, the man is completely blind although his eyes and their nerves are uninjured. Similarly, if a person has a tumour growing in his temporal lobes, he is liable to hear sounds and voices (auditory hallucinations), a state of things not uncommon in the insane. Finally, if by disease his temporal lobes are destroyed, he will thenceforth be permanently deaf.

When the human brain has been mapped out into its motor and sensory areas, there are left over some comparatively large regions which seem to be neither one nor other.

They are the " association centres ", largest in the human brain, smaller in the higher mammals, and non-existent in still lower types. These are the last parts of the brain to assume their functional activity. They are interesting to the psychologist because they constitute the basis of thought, emotion, volition, speech and memory. They are neither sensory nor motor but higher than both ; they act as co-ordinating centres between these others, receiving impulses from the sensory and emitting them to the motor.

The speech-centres, which are very large in man, are the most important of our association centres. Speech is an intellectual activity which requires an elaborate material substratum. It requires centres in which the memories of both heard and seen words are stored ; these inform the motor speech-centre which in its turn innervates the motor areas (Ferrier's) which finally actuate the muscular mechanisms whereby speech is made articulate.

Thus the cerebrum is the organ of the mind, the physical basis of personality. It is the only organ to which we find consciousness always related. This relationship is remarkable; for whereas it would seem that every conscious state is dependent upon an underlying cerebral condition, yet it appears to be a fact that there can be cerebral activity without the arousing of consciousness. "Unconscious cerebration" would seem to be a case in point. And further, a multitude of experiences can be stored in the potentiality of the subconscious mind.

The modifications of the cerebral substance necessary for the enregistering of conscious experiences have been conveniently called "engrams". These are the permanent material differentiations in the neurons of the cerebrum which are specific and essential for each memory. The longer they have been laid down, the more deeply are they engraved, so that the old man who forgets the events of yesterday can relate to you with perfect detail what happened in his boyhood, for—

Time but the impression stronger makes,
As streams their channels deeper wear.

CHAPTER XIV

WHAT THE MICROSCOPE HAS ACHIEVED FOR MANKIND

MOST people have a very hazy idea of the vast extension of knowledge for which science is indebted to the microscope. It should be interesting to recount just what that instrument has enabled us to understand.

In the first place, by the aid of the microscope it was established that a living thing always proceeds from a living thing; that life invariably comes from life.

In all those cases where living things have apparently arisen in liquids without other living things preceding them, the microscope has shown that some—no matter how few—living things were there before. The problem of the arising of life from non-life had narrowed itself down to such a case as this: if a putrescible fluid, broth or milk, be allowed to stand uncovered for a few days, a drop of it viewed under the microscope will be seen to be swarming with minute vegetable organisms popularly called the "germs" of decomposition; but if a similar quantity of the liquid has been boiled and at once protected from the air, then no living things will be found to be growing in the liquid.

In all cases where an observer had believed that he had seen living things which had arisen from non-living material, the microscope was able to show that he had not previously destroyed all the life originally in the liquid, or that he had subsequently allowed the entrance of dust into his "sterilized" liquid. The microscope is always the ultimate court of appeal in such cases, and its verdict is invariably—no life except from life!

This does not mean that at some time in the past, life may not have originated from the non-living, or that in the future some man of science may not be able to bring about this emergence, but it asserts as a matter of fact that in the experience of such men life has never so arisen. Technically put, it is that in our experience there is always biogenesis, never *a*biogenesis.

The next age-long problem which the

microscope solved was the exact mechanism of reproduction. This had been the most tantalizing mystery to all thinkers from Aristotle to Harvey. How and by what means did the young creature or embryo arise? In the body of the female, undoubtedly; but what was the essential part played by the male parent? The microscope showed that the precursor of the next generation was a minute ovum or egg, and that this, unless penetrated by a still more minute male element or sperm, would not develop at all. The rendering visible of the microscopic ovum and of the still more microscopic sperm gave the clue to the whole mystery. It was a Russian physiologist, von Baer, who first saw the living ovum of a mammal (rabbit), just about one hundred years ago.

LIVING TISSUES

The microscope has been of incalculable importance to medicine and to the healing art in that it has given us the conception of the "living tissues". A "tissue" is one of the essential constituents of the body; thus bone, muscle, skin, nerve, and gland are some of the tissues.

Even before these were studied under the microscope we had indeed an idea that they were not all of the same nature because, for one thing, each looked very different from the other even to the unaided eye, but the lens showed in extraordinary detail in what way precisely each tissue differed from the others.

Thus, for instance, muscle was seen to be composed of long, semi-fluid cylinders, nerves of very long, narrow fibres, bone of a most complicated system of canals, skin of many layers of invisible " cells ", and so on. The histologist, as he is called, is able to describe the pattern or system on which each tissue is built up, so that in a few moments, armed with his lenses, he is able to tell whether he is looking at a minute scrap of tooth, bone, brain, muscle, skin or gland.

This knowledge of the ultimate constitution of the tissues of the living animal body is the basis of the new microscopic biology, and of entire departments of medicine whose very names were not invented a few years ago.

Further, it became possible to say exactly in what particulars diseased tissues differ from healthy. Disease, for the most part, has to be long established before the naked eye can recognize it, but by means of the microscope the very earliest stages or phases of disease are revealed. There have in this way been gratifying advances in scientific medicine.

But the microscope did further service to medicine and surgery, for it made it quite clear that such a thing, for instance, as a tumour, was not something composed of a kind of tissue previously unknown but was of the same essential pattern as one of the five or six familiar tissues. A tumour of bone, of muscle, or of gland was merely bone or muscle or gland growing excessively where it ought not so to be growing. It was "exceeding the local" (growth) "limits".

Similarly the microscope revealed a most characteristic plan in the tissues of plants, so that the new science of Microscopic Botany was created. We now know that the invisible structure of a leaf is quite different from that of a stem and from that of a root. The origins of the diseases of plants have similarly been made plain; we understand how to attack the smuts, "rusts" and other parasites of our trees and shrubs.

The crowning achievement of microscopy was the discovery that most infectious diseases are produced by the growth in the body of vast numbers of extremely minute vegetable organisms, each giving rise to a distinct kind of disease. Thus it was made clear that, for instance, diphtheria, tetanus, rabies, anthrax, typhoid fever, plague and tuberculosis were each due to a different kind of germ identifiable by microscopists.

The microscope showed that all such processes as the fermentation of sugar, the souring of milk, the decomposition of a dead body and a large number of the infectious diseases were alike due to the activity of myriads of minute vegetable organisms allied to the fungi. This is sometimes still called the "germ *theory* of disease"; but it has long passed from the realm of conjecture to that of established fact. Only a few ineducable "cranks" still refuse to admit the truth of it.

The microscope has provided the clue as to why so often in the past wounds became putrid and blood-poisoning supervened. Pasteur had shown that milk became sour because minute vegetable organisms had begun to grow in it. Lister, therefore, taking this as a starting point, scrutinized the discharges of wounds, and did, indeed, find in them similar vegetable organisms.

By much patient research he conclusively proved that if all microscopic living matter is excluded from a wound, it will heal without inflammation or putrid discharge. Lord Lister, amid apathy and ridicule, silently and surely transformed surgery from a dangerous art to a safe science.

Finally, the microscope has also been used in the realm of the *non*-living. It has revealed the structure of rocks, precious stones, minerals, and metals. By its aid it has been shown that in a wire over-stretched or over-twisted, the invisible structure has undergone alteration, and this alteration or "fatigue" of the wire has actually been demonstrated under the microscope.

This instrument, the microscope, has introduced us to another universe—that of the infinitely little. For just as the telescope is our introduction to the universe of the infinitely great, after all only a part of creation, so the microscope has been our avenue of entry into a world as abundantly peopled, as wonderful and as incomprehensible as the world of the starry heavens. And as regards mankind this world of the Invisible is far more important for his happiness and far more significant for his well-being.

CHAPTER XV

HOW MANY TASTES HAVE WE?

THE number of true or pure tastes which we possess is smaller than most people think. We have, in fact, only five true tastes (gustations), namely: Acid (or sour), alkaline (or soapy), sweet, bitter and saline (salt).

Vinegar is sour, baking powder is soapy, sugar is sweet, quinine is bitter and salt is saline. Anything in addition to these that we may perceive in regard to things we take into the mouth belongs either to the group of the odours, to that of contacts, or to that of sensations of heat and cold. In other words, nerves of taste, of smell, of contact (touch) and of heat and cold are all involved in what we call " tasting " anything.

An example may make this clearer. If you are asked what the taste of an apple is like you might reply, " Oh, well, just its own taste, the taste of an apple". But there is no such taste; an apple must be sweet or sour; what we call the "taste" of the apple is its flavour or smell, which can be perceived before the fruit is taken into the mouth at all. The so-called "taste" of the apple is an odour, something volatile, which ascends from the mouth to the nose by the back of the throat and is there perceived through the nerve of smell.

Several sensations are aroused by the eating of an apple. There is the contact with its solid substance, which is mediated by the nerves of touch or contact in the tongue, gums, cheeks and palate. Then if the apple is ripe there is the sensation of sweetness; then the sensation of cold, especially if the fruit has been in cold storage, and, lastly, there is the smell, or odour of it popularly called "the taste".

All those four kinds of nerves of sensation are generally involved in the "tasting" of our ordinary food. Now one set in this group, the olfactory, can be voluntarily abolished, as when we close the nostrils while eating the apple, in which case no air ascends from the mouth to the nose, so that we cannot smell the apple and therefore cannot recognize it as "apple". We derive in this case very little so-called "gastronomic" (better olfactory) pleasure from it.

Anyone blindfolded and told to "hold his nose" has the greatest difficulty in distinguishing a raw (if not too sweet) apple from a raw potato. The writer has seen a person thus quite unable to say which was which. A bad cold-in-the-head gives us the same experience; we say our "taste" has gone, whereas the five true tastes are still intact, but the nasal congestion, preventing the ascent of air from the mouth, has abolished our sense of smell. All the five true tastes are retained in a cold.

So important are these odours or flavours in the enjoyment of one's food that when people have a cold they say, "Since I cannot taste anything, it doesn't matter what I eat just now." The pleasure we derive from eating is very largely olfactory. Our enjoyment of the flavour of roast meat, cheese, wine, tea, and fruit all depends on our sense of smell.

There is, for instance, virtually no true taste in roast beef; it is not sour, soapy, bitter, sweet or salt; we add salt to make it more "palatable", and we don't eat beef for the sense of contact. We eat it for its odour, the odour of slightly burnt fat (learnedly called empyreuma); were it odourless we should eat it only from a sense of dietetic duty.

Indeed, so many of the staple articles of our food are devoid of tastes and flavours that we add all sorts of flavourings and seasonings to make them more interesting. Take the case of a milk-pudding: if the vanilla or cinnamon has been left out, how utterly insipid the rice or the tapioca becomes! We should get merely the sense of contact and warmth from the tasteless, boiled grains, not even the most hungry child would eat much of it. The secret of good cooking is by flavourings and seasonings to appeal, not to the palate, but to the inside of the nose.

We speak of a person having a delicate "palate", but we really mean he has an acute sense of smell. The attraction of wine for connoisseurs is in its flavour or bouquet; they sip it slowly, they do not take it "in a mug". Liqueurs, in particular, are sipped so that their flavour may be fully appreciated. Those who are very particular about brandy take it from a large goblet, which becomes filled with the volatile ethers and other constituents of the bouquet that give a massive olfactory sensation.

This sort of thing distinguishes the gourmet from the gourmand; the gourmet eats and drinks—it may be sparingly—and chiefly for the olfactory pleasure of the food. The gourmand eats largely without regard to the flavour of it; he is merely satisfying his large appetite with quantity rather than quality. The gourmet is a gastronomic artist, the gourmand a mere stoker of the vital furnace.

Alcohol has no taste in itself; it is neither acid nor alkaline, sweet, bitter or salt.

A sweet wine or cider is sweet from sugar like any other sweet thing. Beer is bitter from hops, an inferior port from tannic acid. The flavours of the various wines are appeals to the olfactory sensations. *Pure* alcohol has an irritating effect; it stimulates neither the taste nor smell nerves, but irritates the contact nerves in the tongue, palate and gullet thus giving rise to the hot sensation once described by Pat in *Punch* as "going down like a torchlight procession".

Some persons who abuse alcohol take it for this irritant effect on throat and gullet; hence these public nuisances will drink anything that is irritant, such as methylated spirit, or even Condy's fluid. Some drunkards, of course, drink alcohol for the intoxicating subjective effect, and these will drink ether because, being so rapidly absorbed, it intoxicates very quickly. We call those people "nuisances" because, on account of them, other people who know how to use alcohol are worried in a hundred ways.

Exactly the same irritant effect on the touch nerves is produced by pepper, mustard, capsicum and the many condiments. We do not really *taste* pepper or mustard, it is the contact nerves that are irritated by these things. Peppermint, for instance, stimulates the cold-perceiving nerves, so that just after it has been taken, a current of air at the ordinary temperature seems quite cold.

Several sorts of nerves are therefore stimulated by an ordinary dinner; the true tastes are all there with the possible exception of the "soapy", and even of this a banana is a very good imitation.

It is noticeable how we add materials to foods in order to provide them with "tastes"; to get sourness we add vinegar, to get sweetness sugar, to get saltness salt, to get bitterness hops, though it is only rarely that we desire this last taste.

Few foods are in themselves sour—butter-milk is an exception—few are salt, few are soapy or bitter. Hence the insipidity of meat or fish taken by itself is natural; and this is the meaning of the addition of so many accessory materials to food—salt, flavours, relishes, condiments, herbs, garnishing, etc.

The true tastes have different "end-organs", or "taste-buds" as they are called, and they are distributed differently as follows: The sweet-perceiving are on the front of the tongue, the acid-perceiving are at the sides, while the bitter-perceiving are at the back. Thus to taste a sweet liquid it is quite enough to put the tip of the tongue into it, it does not need to be swallowed, whereas a bitter substance is not tasted until it has nearly gone over the back of the tongue. If some time has elapsed since we last took quinine, and we are swallowing our dose slowly, we have time to think it is not so bad after all until it reaches the full development of its disagreeableness literally at "the bitter end".

Some things being perceived on the margins of the tongue are said to "set the teeth on edge", although physiologically they have nothing to do with the teeth. Hence the fine metaphor in the book of Ezekiel—"The fathers have eaten sour grapes and the children's teeth are set on edge". Incidentally it is a poetical description of heredity.

We cannot here give all the evidence that the taste-buds differ physiologically in the different parts of the tongue, but we may refer to the evidence from mechanical and from electrical stimulation and to the effects of drugs.

Tapping the front of the tongue with a clean glass rod produces a sweetish taste, and rubbing the rod against the side of the tongue will give rise to a bitter sensation. Alternating (induced) electric currents, too feeble to affect the retina or the skin, and too weak to cause electrolysis, can produce distinctly different tastes according to the parts of the tongue stimulated.

Lastly, by chewing the leaves of the African plant, *Gymnema sylvestris* (one of the *Asclepiadaceae*), only the bitter-perceiving end-organs are quickly paralyzed, so that a very bitter medicine may be comfortably swallowed. Later on, the sweet taste is also abolished by gymnema.

Cocaine, which in high concentrations

abolishes all the five tastes, as well as touch, temperature-sensations and pain, abolishes the bitter tastes first.

Taste-buds are found only in the mouth, not, for instance, in the gullet or the stomach, so that we cannot taste food when once it has been swallowed. The pleasant taste or odour of food or drink may tempt one to take more of it than the appetite would warrant, and therefore indigestion may be brought on in this way. The appetite is sated before the sensation of smell is.

We appreciate some things for the massive, contact sensation, as, for instance, the thick liqueurs with their non-irritant oily contact.

When, as children, we had to take castor oil, it was not its oily contact we so disliked but its nauseating odour, and so in the nursery we were told "to hold" our noses or had them held vicariously for us. This prevented the odour, not the taste, getting to the nose; for all kinds of castor oil are virtually "tasteless". Sometimes the odour of the oil was masked or antagonized by that of orange juice or sherry, a process similar to that of gilding the pill.

It will be apparent that we can perceive two different tastes at the same time, in fact we can taste two such opposites as bitter and sweet, for we can mix quinine and sugar in the right proportions to stimulate simultaneously the bitter- and the sweet-perceiving end-organs. Lemonade is acid and sweet at the same time; a ripe banana is sweet and soapy at the same time.

Many things other than sugar are sweet —glycerine, saccharine (which is hundreds of times sweeter), and acetate of lead, which is so sweet it is called "sugar of lead". Now the chemical constitutions of these substances have hardly anything in common; we do not know why these things, so different from one another, can all stimulate the sweet-perceiving nerves. Similarly, many things very different chemically are all bitter such as quinine, strychnine, poisonous cheese, unwholesome mushrooms, Epsom salts and bitter almonds. And, of course, the number of different acids known to chemists is enormous, but they all stimulate the same gustatory nerves.

The taste nerves seem to be paralyzed by extremes of heat and cold. Something that may be taken very hot, as, for instance, tea, can easily prevent our getting the flavour of the beverage. Similarly, very cold viands can render the "palate" insensitive to the taste of something taken immediately afterwards, just as great cold renders the gums insensitive to the pain of having a tooth pulled out.

Ginger, besides irritating the contact nerves, also irritates the heat-perceiving nerves to such a degree that the hot sensation long outlasts the stimulus of the ginger; thus "ginger is hot i' the mouth". It also results from this that the tastes of viands taken after ginger or peppermint (which stimulates the cold nerves) are imperfectly perceived.

This sort of thing is really due to contrast in tastes and odours, and hence we know that some things taste well, others badly, after certain tastes and flavours which linger on. Connoisseurs study these things; they do not eat garlic just before a chocolate *éclair*, or an onion before a meringue. The scientific choice of courses and dishes consists in selecting them in such an order that the tastes and flavours shall follow each other pleasantly and without harsh contrasts. We should not go from *crême de menthe* to fried fish, the contrast is disagreeably sudden.

The "smell of cooking", that of hot grease, is to many people so unpleasant that they cannot enjoy the pleasures of the table which, being very largely olfactory, are liable to be overpowered by it. We do not enjoy strawberries and cream in an atmosphere redolent of hot broth. Some few people have survived to this day who do not like to take their food in an atmosphere laden with tobacco smoke. They maintain, correctly, that if their nostrils are full of smoke they cannot perceive the delicate perfumes of certain foods and beverages. Hence having to take a meal in an atmosphere laden with tobacco smoke is to those people an aesthetic outrage.

The flavour of good tea, for instance, cannot be perceived, and therefore enjoyed, when one's nostrils are full of smoke. Smokers, who are notoriously considerate, evidently cannot be aware of the physiology and psychology of the five true tastes.

CHAPTER XVI

THE CAUSE OF COLOUR

COLOURS are made by light, and light—even that of the moon—comes from the sun.

Sir Isaac Newton discovered that when a beam of light was passed through a solid wedge of glass (a prism) it was broken up into a band of seven colours—red, orange, yellow, green, blue, violet and indigo. The band, which is usually so arranged as to have red at the left-hand side, is known as "the spectrum." The rainbow is a spectrum on a gigantic scale in the clouds.

White light, then, is broken up into these seven spectral colours by being bent (refracted) on going through a prism. The colours are potentially present in white light, although the eye cannot detect them. Newton also did the converse experiment of recombining the seven colours by means of a lens into white light again.

Now the colours of objects depend on how the materials of these objects deal with the light that falls on them. If an object sends on to our eyes all the rays that fall on it, it is white; snow does this, a piece of chalk and white writing-paper do it.

We possess an instrument whereby we can tell whether all or some or none of the coloured rays in the light are coming from any given object; it is an arrangement of prisms known as the spectroscope. This breaks up light into the seven colours, but if one of these colours or even a small portion of it is not present in the light, then there is a black band or line in what we see in the spectroscope, and we call that an "absorption" band or line.

When we look at the light coming from snow or from chalk, we find all the seven colours in the spectrum as presented to our eyes by the spectroscope. When, however, we look at a black object with the spectroscope we can see no spectrum at all, for it has absorbed all the light that fell on it and absorbed all of it equally. There is no light left to come on to the spectroscope.

Suppose, now, that we have our band of colours in a dark room, and, taking a piece of red flannel, pass it along the series from red to indigo, it will look red in the red but black in all the others. Evidently, the red flannel is "red" because in white light it quenches all the colours except red, and so allows only that colour to affect our eye. In Nature no flower is perfectly black, because no flower absorbs all the rays of light which fall on it.

Although there are only seven colours in the spectrum, there are thousands of "colours" both in Nature and in the arts. Some "colours" are really dilutions of a pure colour with white light, as, for instance, pink which is red plus white, "eau de Nil", which is green plus white, and so on, we call these "tints".

Then, again, there are the "shades", which are mixtures of black with a particular colour—the so-called dull or "sad" colours —for instance, red plus black which makes a dark red or a brown.

Further, many colours not in the spectrum are formed by the mixing of two or more which are; and these mixed or compound colours may again be diluted with white or mixed with black. Thus purple is a mixture of red and blue, and it may be obtained as a light purple or a dark purple.

CHAPTER XVII

SUSPENDED ANIMATION

THERE is a state into which living things can pass which is a kind of intermediate state between life and death. It is also known as "suspended animation" or "latent life". The living matter (protoplasm) in this state does not show any of the familiar signs of life, and yet it is not dead, because by appropriate means it can be revived and made to show these signs again.

The signs referred to are: (1) the taking in of oxygen, (2) the giving out of carbon dioxide (sometimes called carbonic acid gas) and water, and (3) the production of animal heat. All living things carry out these processes with greater or less intensity. The carbon and hydrogen of the tissues are oxidized by the oxygen absorbed, and heat is the inevitable result of this oxidation. There are other signs of life which for our present purpose we need not examine.

Luckily these three evidences of life are all capable of being measured, for by appropriate apparatus we can estimate the amount of oxygen taken in by the body, the amounts of carbon dioxide and water produced, and the quantity of heat set free in a given time.

There is a further sign of life, the production of an electric current, and this also is measurable, although it requires the use of a very delicate and costly apparatus, the galvanometer.

So long, then, as an animal is alive it is always absorbing oxygen, excreting carbon dioxide and water, and producing some heat and electricity; if it is doing none of these things, we are entitled to regard it as dead.

Now, excluding the electric sign in the meantime, we find when we come to test certain organisms at certain times that they are giving no evidences of vitality as just described.

Let us begin with some dry seeds in a pill-box; they are evidently not dead, because, we know, if sown in moist earth they will grow into plants, but all the same they are not showing any of the three signs of life.

The ordinary person could not tell by

looking at them whether they were alive or dead ; but if they are able to grow into plants they must be alive. And yet if they are not taking in oxygen or giving out carbonic acid gas they cannot be regarded as *wholly* alive ; therefore we are compelled to regard them as in an intermediate state which we name " latent " life. Latent life is biologically purgatory without its punishments.

This intermediate state the Germans call *Scheintod* or apparent death, which is the other aspect of the same fact.

But it can be shown that these so inert seeds, if they be given a sufficiently powerful electric shock, will respond by producing an electric current in the galvanometer, so that this latest discovered sign of life is also the most infallible ; it will be given when the other three will not.

It is well known that seeds can for many years remain capable of germinating, although the tales of being able to grow wheat from seeds found in Egyptian mummies are now known to have been Arab deceptions.

Latent life in the animal kingdom is a particularly interesting condition. Desiccation or drying of certain lowly animals will send them into a state of vital latency just as it did the seeds. Ever since the discovery of dried-up wheel-animalcules (Rotifera) by the Dutch naturalist Leeuwenhoek in 1719, we have known that certain animal organisms could enter upon an indefinitely long period of suspended animation. Leeuwenhoek found that those little creatures could remain in dried mud for a very long time without showing any movement, but that they became active on being moistened.

Other organisms which can withstand prolonged drying are "bear animalcules" and "paste-eels" and some of the very simple group of the Infusoria.

Among the cold-blooded animals there are many cases of latent life, for snails, water-beetles, frogs, and fish can all be frozen and yet remain alive.

All life tends to become less active as the temperature is lowered, until at a certain reduction the vital manifestations cease, but even then it does not follow that the life is extinguished.

Sir John Franklin, in his polar expedition of 1820, reported finding carp fish frozen so solid that the intestines of some of them could be taken out *en masse*, and yet on being thawed before a fire some of their companions "revived and moved about actively". A batch of fishes, frozen in a block of ice at minus 15° C., have been known to revive on being thawed, although the bodies of some of the others were so hard that they could be powdered up along with the ice.

A fish has been frozen in a block of ice, then sawn in two along with the ice, and each half, when the ice was melted, performed active movements before it died.

Sir Ernest Shackleton stated that in the south polar seas there are organisms frozen for ten months of the year, and active only during the other two. This is the sort of thing we mean by Latent Life.

A frog can be frozen solid and brought into the lecture-room so stiff that it can be held out by the toes like a stick; and yet after being thawed during the lecture, it will hop about as though nothing had happened. Experiments by the author have shown that so long as ice does not form around the heart, frozen frogs will remain alive.

The hibernation or winter sleep of animals is a state closely allied to latent life.

There is no doubt that the amphibia and the reptiles (snakes, tortoises, etc.) can enter upon a prolonged period of depressed vitality owing to the low temperatures they can endure. As frogs bury themselves in the mud under the ponds and lakes where they pass the winter, it is not probable that they are actually frozen all that time, but they are hibernating.

In the hibernation of the warm-blooded animals, we have a state closely resembling true latent life.

Bears, dormice, marmosets, hedgehogs, and other animals, having laid on plenty of fat against the approach of winter, seek out hiding-places, and retire from the world until the next spring. In those cases of warm-blooded hibernation, though the animal is not breathing, its heart is beating, and its gaseous exchanges are reduced to a minimum. It is living on its fat which it oxidizes for heat, and therefore emerges in the spring very thin and cold.

It need not be pointed out that the warm-blooded animals, by reason of the much greater delicacy of their bio-chemical processes, could not survive the loss of water or the loss of heat which more lowly creatures can endure.

When we come to man himself, we find from time to time cases of such a nature occurring that we can only regard them as closely allied to hibernation and latent life; popularly they are called " trance ", learnedly narcolepsy (a falling into sleep).

Every now and again we hear of persons who fall into a deep sleep or coma in which state they remain for days or weeks, scarcely breathing and with the very slightest action of the heart. The pulse at the wrist may actually be imperceptible, and the heart sounds inaudible.

One of the most famous cases is that of Colonel Townsend, reported on by the well-known Dr Cheyne of Dublin. The unusual feature in this case was that Colonel Townsend could go voluntarily into this state of suspended animation, remaining breathless and pulseless for many hours.

Most certainly this is the nature of the condition of those Fakirs in India who are enabled to undergo such extraordinary feats of endurance announced from time to time in the Press. Some of these are too well authenticated to be attributed to fraud or explained by collective hallucination. A man has allowed himself to have his ears and nostrils stuffed with cotton-wool, to be sewn up in a sack, which was placed in a locked chest, and then be buried for periods as long as six weeks, at the end of which time the doctor who saw the man buried proclaimed him alive.

Now and again persons have been prepared for burial who were not dead but only in this state of narcolepsy; such cases, however, are extremely rare in spite of sensational representations of them, such as those in the famous Musée Wierz in Brussels.

Shakespeare, who has noticed almost everything, has remarked also on this state of suspended animation, for in *King Lear* (v. 3) the king says—

> I know when one is dead and when one lives;
> She's dead as earth. Lend me a looking-glass;
> If that her breath will mist or stain the stone,
> Why, then she lives.

Death is a state of protoplasmic immobility which cannot be recovered from, a state, as it has been called, of "infinite functional inertia". Latent life and not sleep is the "image of death", for just as we awake from sleep, so from latent life there can be a protoplasmic resurrection learnedly called "Anabiosis". In life "the sands of time" are running out rapidly; in latent life the stream has been mysteriously arrested; in death the sand is all in the lower globe never to leave it.

CHAPTER XVIII

SOME LOST ARTS

It would seem an incontrovertible statement that in this age, when science is applied to everything, we should know how to do everything. But do we? Are there no lost arts?

An example that of course occurs to everyone is the building of the pyramids by the ancient Egyptians. Our engineers could pile huge blocks of stone one upon another over a central chamber, but they would do it by steam-driven machinery. We who have made the useful barrages over the Nile could also rear the useless pyramids if we wished. The Egyptians may have employed slave labour and used levers, inclined planes and rollers, but the fact is that we do not know with certainty how they caused these gigantic blocks of granite to rise into the air of the desert.

Passing on to the Middle Ages, what shall

we say of stained glass? We are making coloured glass for churches every day, but antiquarian experts tell us that the glass we make now lacks some quality or attribute which the old glass possessed. Hence it is, for instance, that the pre-Reformation glass is so highly prized by those who have specially studied this subject. We certainly know a great deal about pigments and a great deal about glass, but we have lost the art of combining these as they were combined in the stained glass of long ago so that it came to be a thing of beauty in the haunts of ancient peace.

In the making of varnish we have another instance of a lost art. This is proved by the following quotation from an article which appeared lately in *The Times*:—"Recently the rector of one of Wren's most famous churches was surprised to receive a visit from three varnish experts, two American and one English, who requested permission to examine the varnish in the church. They declared that the oak panels in the famous vestry of the church represented a lost secret in the making of varnish. The varnish employed in the seventeenth century served as a preservative of the wood, but was so transparent that it did not obscure the beauty of the grain. The ingredients which resulted in this quality are not known ". Comment is unnecessary.

Road-making may be taken as our next example. No doubt at the present day our highways are constructed on the soundest mechanical principles; but for hundreds of years the roads in Great Britain were nothing less than scandalous makeshifts. The art of road-making was well known to the Romans. Roads from one end to the other of their vast empire were made in the first instance for military purposes, and were therefore made to last. They carried the heaviest traffic of their time, and they were properly drained. In making them the Romans first dug out the soil to a depth of several feet, and then laid down stones in layers, the largest below, the smallest on the top. The small surface stones were bound together by a top-dressing of soil. Where the road ran through camps or between buildings, the surface seems to have been formed of flat stones placed side by side after the manner of a pavement; the ruts formed by the chariot wheels in such roads can still be seen at Pompeii, and at one or two particularly well preserved camps in England.

After the Romans left Britain in the fifth century, no attempt seems to have been made to keep these highways in repair or to make new ones on Roman principles. Most of the roads in England were mere tracks over the natural surface of the ground, which became quagmires in wet weather and dusty hollows in dry. Where the traffic was concentrated on the main route leading into or out of a town, the road became worn into a deep depression naturally called " the hollow way ". Hence the frequence with which we meet this as the name of a street in many places in England. Holloway Road in London is perhaps the best known of these, but many English towns also have their " Holloways ", pointing back to the time when the traffic had worn the road into a deeper and deeper depression. These undrained, unmade roads of the England of the Middle Ages had no " bed " and no foundation ; the art of road-making had been lost. The roads, in fact, were so bad that travelling was done almost entirely on horseback, and this undoubtedly delayed the development of wheeled traffic. When the stage-coach did come, it was constantly sticking in the mud, as was the huge carriage or "family" coach in which the wealthy moved from one part of the country to the other, for the rest of the people scarcely travelled at all. This was the highwayman's opportunity; he appeared, pistol in hand, at the window of the embarrassed vehicle, and had things pretty much his own way. Well on into the time of the Georges the the roads of Great Britain were a disgrace to a civilized country.

From this slough of despond they were raised by the energy of a Scotsman, one John Loudon MacAdam, a native of "the auld toun o' Ayr". MacAdam, whose name of course survives in "macadamized", was born in Ayr in 1756, and died in 1836 in the little watering-place of Moffat amongst the Dumfriesshire hills. MacAdam reintroduced the Roman method of building up the road from a bed of stones of diminishing size. After making experiments on his estate in Scotland, he published in 1819 his, "Practical Essay on the Scientific Repair and Preservation of Roads". In 1827 MacAdam was appointed General Surveyor of Roads, and the rest of his life he spent almost entirely in England. He declined a knighthood.

The only roads of proper construction in Scotland prior to the macadamized roads were those made in the Highlands by General Wade after Culloden. They are the main roads over the Grampians to-day. Wade's engineers did their work skilfully, building most substantial highways with not too steep gradients or dangerous curves. The admiration of someone who had more Hibernian wit than poetry in his constitution is embalmed in the well-known lines—

If you had seen these roads before they were made,
You would lift up your hands and bless General Wade.

Our next example of forgotten arts may be taken from the insanitary condition of the mediæval cities of Europe. Imperial Rome brought pure water for drinking purposes from across the Campagna in a noble aqueduct; and she contrived the opposite of this in a system of sewers the best known of which is the *Cloaca Maxima*. We are so accustomed at the present moment to our cities having plentiful supplies of pure water and possessing excellent sewerage systems that we readily imagine it has always been so. But this is very far from the truth, for the mediæval cities of Europe were very poorly supplied with drinking water, and had practically no sewerage at all: these were two lost arts.

As late as the time of Charles II, the streets of London were in a disgraceful condition; heaps of decomposing garbage polluted the air. The continual outbursts of pestilence were directly due to the insanitary state of these streets which were never clear of decaying vegetable and animal matter. Old Edinburgh was, if possible, in a worse state. Here within the wall built after Flodden, the houses—unable to spread in any but the vertical direction—rose to the dizzy height of ten or twelve storeys. These towering and highly picturesque "lands" had no plumbing whatever. The drinking-water obtained from wells far down in the street below was carried in barrels on the backs of porters or "water-caddies", as they were called, to the top of these immensely high houses. The slop water was disposed of by a "gravity system" conspicuous in its simplicity, for it was merely thrown out of the window on to the pavement below. Hence the ominous cry of "Gardey loo!" a Scottish corruption of *Gardez l'eau*, "beware of the water". It might happen that the lady in her satins and laces, who had just alighted from her sedan chair and who had not looked above, might be splashed by the malodorous liquid; while we may be sure that it was the constant prayer of the bewigged male *not* to "let some drops descend on me". Smollett, who knew Edinburgh well, makes one of the characters in *Humphrey Clinker* explain the phrase "gardey loo" as meaning, "May the Lord have mercy on you". "Auld Reekie, I can smell ye noo", was the familiar exclamation of the traveller from the south on approaching the venerable city if the wind happened to be in the north.

Of baths in mediæval Europe there were none; baths had vanished with the Roman Empire. One of the grievances the Christian Spaniards had against the Moors in Spain was that the Moors took baths. The baths of ancient Rome are amongst the most conspicuous of her ruins. Of course here and there in the Renaissance palaces of Italian princes bath-rooms might have been found; Titian decorated one in the Vatican. Except on the American continent, even at the present moment the bath has not been restored to its proper place in the house. Public baths have been instituted well within the memory of some of us; only the wealthy and leisured few take Turkish baths regularly even now. The spacious times of Queen Elizabeth were not spacious enough to include bathing.

Another lost art is that of heating economically the ordinary dwelling-house. Our methods are miserably inferior to those employed by the Romans centuries before the birth of Christ. No doubt we *can* heat our large buildings, mansions, museums, and churches so as to make them more or less tolerable, but the ordinary house of the ordinary person is far more uncomfortable in an English winter than was a Roman villa here two thousand years ago.

The Romans heated their houses by heating first the floors and the walls, when all was well, calorifically speaking, with the house. We have substituted the open fire which, while it ventilates well, heats very badly, besides being uneconomical and dirty. In the hypocausts of the one storeyed Roman buildings the hot air passed continuously under the floors, and in specially constructed flues through the walls. The same method of heating was used for the water of the baths, a bath being almost always found amongst the ruins of a Roman villa. If our feet are warm, we are warm altogether; if cold, we are cold. The heating of floors, then, is almost a lost art. I say "almost", because it is a fact that a few large buildings in London have quite recently been heated by a system of hot-water pipes laid down within the floor and in the walls and ceilings of the buildings. Before we leave the topic of Roman construction we might remind ourselves that some architects believe that the secret of the admitted excellence of Roman mortar is lost.

There is little doubt that one age forgets what an earlier age knew. Surgery under the Romans reached a degree of excellence which it was not again to attain until the time of Ambrose Paré, the Father of Surgery, who died in 1590. Paré by reintroducing the ligation of blood-vessels which had been in abeyance since the time of Celsus—the reign of Tiberius Cæsar—made the technique of the amputation of limbs practically what it is to-day.

No fewer than two hundred kinds of surgical instruments were found at Pompeii—a far larger number than throughout the entire Middle Ages was considered necessary for a whole college of surgeons. Such operations as herniotomy, the Cæsarian section, removal of the lens for cataract, and those in plastic surgery were all quite familiar to the Romans in the first century of the Christian era.

The eminent medical historian, Dr Fielding Garrison of New York, assures us that from the death of Soranus of Ephesus in the second century A.D. there were no real additions to obstetrics for 1,500 years. Almost everyone who knows anything about medical lore knows that what Galen taught in the second century was the orthodox belief in the medical schools of Europe for 1,400 years; but it is possibly not so well known that a great deal that Galen knew was completely forgotten by these same schools. Galen knew of more than one form of phthisis; he knew of its infectivity, and that a sea-voyage or residence in high and dry altitudes was very beneficial for it. Its infectious character has been virtually re-discovered in our own day. From the death of Galen to the time of Harvey, 1600, there was practically no experimental physiology.

Let our last example be that of the Sun-Cure. Modern medicine is preaching the great value to health of plenty of sunlight.

To the ancients this would have sounded the tritest of truisms. There have been sun-worshippers since the dawn of terrestrial intelligence. The Romans, though providing for shade in the courtyards of their dwellings, selected the sunniest spots, at least in Britain, for their villas. The great religious houses of the Middle Ages were built wherever possible on the open sunny sites so much admired to this day. Think of the abbeys at Fountains and at Tintern. With the castles came in the reign of gloom.

The walled cities of Europe fostered the clustering of houses so close together that from a vast number of the narrow streets the sun was completely excluded. Then came even the worse slums of the towns of the so-called Industrial Revolution, the very high places of all that was contrary to health. To make matters worse, there was actually a tax on the area of glass in the windows of houses which, reducing windows to the smallest possible size, effectually produced the greatest *un*happiness of the greatest number.

Our great-grandfathers never seemed to have realized that the sunlight which they took for granted was in itself a positive factor in the upbuilding and maintaining of a healthy body. And so we had sunless houses, sunless schools, sunless factories, sunless churches and sunless streets, and—tuberculosis, rickets and anæmia. Our forefathers, accepting the sun as a gift to the just and to the unjust, neglected to see that every one of these might really enjoy it. We are now proving tediously in the laboratory that in the free and universal sunshine we have one of the most valuable positive factors in the fostering of national health.

For Product Safety Concerns and Information please contact our EU representative GPSR@taylorandfrancis.com Taylor & Francis Verlag GmbH, Kaufingerstraße 24, 80331 München, Germany

Batch number: 10725863

Printed by Printforce, the Netherlands